W9-CYG-601

Air Quality in Urban Environments

ISSUES IN ENVIRONMENTAL SCIENCE AND TECHNOLOGY

EDITORS:

R.E. Hester, University of York, UK
R.M. Harrison, University of Birmingham, UK

EDITORIAL ADVISORY BOARD:

Sir Geoffrey Allen, Executive Advisor to Kobe Steel Ltd, UK, **P. Crutzen**, Max-Planek-Institut für Chemic, Germany, **S.J. de Mora**, Plymouth Marine Laboratory, UK, **G. Eduljee**, SITA, UK, **L. Heathwaite**, Lancaster Environment Centre, UK, **J.E. Harries**, Imperial College of Science, Technology and Medicine, UK, **S. Holgate**, University of Southampton, UK, **P.K. Hopke**, Clarkson University, USA, **Sir John Houghton**, Meteorological Office, UK, **P. Leinster**, Environment Agency, UK, **J. Lester**, Imperial College of Science, Technology and Medicine, UK, **P.S. Liss**, School of Environmental Sciences, University of East Anglia, UK, **D. Mackay**, Trent University, Canada, **A. Proctor**, Food Science Department, University of Arkansas, USA, **D. Taylor**, AstraZeneca plc, UK, **J. Vincent**, School of Public Health, University of Michigan, USA

How to obtain future titles on publication

A subscription is available for this series. This will bring delivery of each new volume immediately on publication and also provide you with online access to each title via the Internet. For further information visit http://www.rsc.org/Publishing/Books/issues or write to the address below.

For further information please contact:
Sales and Customer Care, Royal Society of Chemistry, Thomas Graham House, Science Park, Milton Road, Cambridge, CB4 0WF, UK
Telephone: +44 (0)1223 432360, Fax: +44 (0)1223 426017, Email: sales@rsc.org

ISSUES IN ENVIRONMENTAL SCIENCE AND TECHNOLOGY

EDITORS: R.E. HESTER AND R.M. HARRISON

28
Air Quality in Urban Environments

RSCPublishing

Lake Superior College Library

ISBN: 978-1-84755-907-4
ISSN: 1350-7583

A catalogue record for this book is available from the British Library

© Royal Society of Chemistry 2009

All rights reserved

Apart from fair dealing for the purposes of research for non-commercial purposes or for private study, criticism or review, as permitted under the Copyright, Designs and Patents Act 1988 and the Copyright and Related Rights Regulations 2003, this publication may not be reproduced, stored or transmitted, in any form or by any means, without the prior permission in writing of The Royal Society of Chemistry or the copyright owner, or in the case of reproduction in accordance with the terms of licences issued by the Copyright Licensing Agency in the UK, or in accordance with the terms of the licences issued by the appropriate Reproduction Rights Organization outside the UK. Enquiries concerning reproduction outside the terms stated here should be sent to The Royal Society of Chemistry at the address printed on this page.

Published by The Royal Society of Chemistry,
Thomas Graham House, Science Park, Milton Road,
Cambridge CB4 0WF, UK

Registered Charity Number 207890

For further information see our web site at www.rsc.org

Lake Superior College Library

Preface

Serious deterioration of urban air quality was initially a consequence of urbanisation itself and the large quantities of wood and coal which were burnt for heating purposes. The situation worsened subsequent to the industrial revolution which caused further large quantities of fuels to be burnt in order to provide the energy needed by manufacturing machinery. Air quality in London throughout the 19th and first half of the 20th century is known to have been extremely poor and particularly so in winter, when larger amounts of fuel were burned, and during weather conditions that inhibited the dispersion of the emissions. Subsequent to the London smog of December 1952, in which it is estimated that at least 4000 premature deaths occurred as a result of air pollution exposure, legislative measures were put into place to clean up the atmosphere. Other developed countries produced a similar policy response and there were dramatic improvements in air quality through the 1960s in particular, which in respect of some pollutants continue to the present day. However, concurrent with this reduction in pollutant emissions from home heating and industry, there was a massive growth in road traffic and consequently in vehicle-emitted pollutants. Only subsequently, in the US from the 1970s and in Europe from the 1980s, were substantial reductions required in the emissions from each new vehicle so as to limit the detrimental effect on urban air quality. A modern vehicle now emits a small fraction of the quantity of pollutants emitted by vehicles that preceded these technological developments, which were themselves enforced by legislation. However, despite the great improvements of recent years, there are substantial residual urban air quality problems and many thousands of people die prematurely each year as a consequence of air pollutant exposure.

This volume is designed to provide an overview of the most important aspects of this field of science. In the first chapter, Ole Hertel and Michael Goodsite set the scene by reviewing the major air pollutants and the typical urban air pollution climates encountered across the world. They also introduce those factors that are important in determining air quality, which are then elaborated upon in subsequent chapters. Jenny Salmond and Ian McKendry

Issues in Environmental Science and Technology, 28
Air Quality in Urban Environments
Edited by R.E. Hester and R.M. Harrison
© Royal Society of Chemistry 2009
Published by the Royal Society of Chemistry, www.rsc.org

consider meteorological factors, which are the key to understanding episodic high concentrations of pollution which can be so damaging to public health. However, airborne concentrations are determined not only by emissions and meteorology but also by atmospheric chemical processes. It is now more than half a century since researchers in California developed an understanding of photochemical smog formed in the urban atmosphere, but research continues in this area and our understanding of even quite basic aspects can be found wanting. William Bloss, in the third chapter, reviews the more important chemical processes involved in determining the composition of urban atmospheres, highlighting some more recent findings.

Having considered the factors determining above-ground air pollutant concentrations, the next chapter by Imre Salma considers air quality in underground railway systems. Because underground trains are themselves a source of pollution, especially airborne particles, and since they operate in a relatively enclosed space, air quality in underground railway systems can be very poor in comparison to the open atmosphere and researching the sources and concentrations of pollutants in underground railways is currently a very topical subject. This leads naturally to a chapter by Sotiris Vardoulakis on indoor and outdoor human exposure to air pollutants. Typically, air quality monitoring uses fixed outdoor sites, but these do not well describe the behaviour of people who are, of course, much more mobile. Such factors are considered in the chapter which considers not only how exposure occurs but also some of the implications of reducing it.

In the sixth chapter, Robert Maynard reviews the health effects of urban air pollution. These include both acute effects which represent rather rapid responses to air pollutant exposures, as well as chronic effects which manifest themselves in reduced life expectancy in people living in more polluted cities. The current state of knowledge and overall implications for human health are considered. In the final chapter, Martin Williams considers the policy response to improving urban air quality. In doing so, he shows how measures taken in the UK have led to air quality improvements and also how new approaches based on exposure reduction will become important within Europe in coming years.

Overall, the volume provides a comprehensive overview of current issues in this long-standing but nonetheless still topical field, which we believe will be of immediate and lasting value, not only to the many practitioners in central and local government, consultancies and industry, but also to environmentalists and policy-makers, as well as to students in environmental science and engineering and management courses.

Ronald E. Hester
Roy M. Harrison

Contents

Issues in Environmental Science and Technology, 28
Air Quality in Urban Environments
Edited by R.E. Hester and R.M. Harrison
© Royal Society of Chemistry 2009
Published by the Royal Society of Chemistry, www.rsc.org

Health Effects of Urban Pollution   **108**
Robert L. Maynard

The Policy Response to Improving Urban Air Quality   **129**
Martin Williams

Editors

Ronald E. Hester, BSc, DSc(London), PhD(Cornell), FRSC, CChem

Ronald E. Hester is now Emeritus Professor of Chemistry in the University of York. He was for short periods a research fellow in Cambridge and an assistant professor at Cornell before being appointed to a lectureship in chemistry in York in 1965. He was a full professor in York from 1983 to 2001. His more than 300 publications are mainly in the area of vibrational spectroscopy, latterly focusing on time-resolved studies of photoreaction intermediates and on biomolecular systems in solution. He is active in environmental chemistry and is a founder member and former chairman of the Environment Group of the Royal Society of Chemistry and editor of 'Industry and the Environment in Perspective' (RSC, 1983) and 'Understanding Our Environment' (RSC, 1986). As a member of the Council of the UK Science and Engineering Research Council and several of its sub-committees, panels and boards, he has been heavily involved in national science policy and administration. He was, from 1991 to 1993, a member of the UK Department of the Environment Advisory Committee on Hazardous Substances and from 1995 to 2000 was a member of the Publications and Information Board of the Royal Society of Chemistry.

Roy M. Harrison, BSc, PhD, DSc(Birmingham), FRSC, CChem, FRMetS, Hon MFPH, Hon FFOM

Roy M. Harrison is Queen Elizabeth II Birmingham Centenary Professor of Environmental Health in the University of Birmingham. He was previously Lecturer in Environmental Sciences at the University of Lancaster and Reader and Director of the Institute of Aerosol Science at the University of Essex. His more than 300 publications are mainly in the field of environmental chemistry, although his current work includes studies of human health impacts of atmospheric pollutants as well as research into the chemistry of pollution phenomena. He is a past Chairman of the Environment Group of the Royal Society of Chemistry for whom he has edited 'Pollution: Causes, Effects and Control' (RSC, 1983; Fourth Edition, 2001) and 'Understanding our Environment: An Introduction to Environmental Chemistry and Pollution' (RSC, Third Edition, 1999). He has a close interest in scientific and policy aspects of air pollution, having been Chairman of the Department of Environment Quality of Urban Air Review Group and the DETR Atmospheric Particles Expert Group. He is currently a member of the DEFRA Air Quality Expert Group, the DEFRA Expert Panel on Air Quality Standards, and the Department of Health Committee on the Medical Effects of Air Pollutants.

Contributors

William Bloss, *Division of Environmental Health and Risk Management, School of Geography, Earth and Environmental Sciences, University of Birmingham, Edgbaston, Birmingham B15 2TT, United Kingdom.*

Michael Evan Goodsite, *Department of Atmospheric Environment, University of Aarhus, P.O. Box 358, Frederiksborgvej 399, 4000 Roskilde, Denmark.*

Ole Hertel, *Department of Atmospheric Environment, National Environmental Research Institute, Denmark, University of Aarhus, P.O. Box 358, Frederiksborgvej 399, 4000 Roskilde, Denmark.*

Robert Maynard, *Health Protection Agency, Centre for Radiation, Chemical and Environmental Hazards Chemical Hazards and Poisons Division (Headquarters), Chilton Didcot, Oxfordshire, OX11 0RQ, United Kingdom.*

I.G. McKendry, *Department of Geography, University of British Columbia, Room 250, 1984 West Mall, Vancouver, B.C., V6T 1Z2, Canada.*

Imre Salma, *Eotvos University, Institute of Chemistry, H-1518 Budapest, P.O. Box 32, Hungary.*

Jennifer Salmond, *School of Geology, Geography and Environmental Science, University of Auckland, Private Bag 92019, Auckland, New Zealand.*

Sotiris Vardoulakis, *Public and Environmental Health Research Unit (PEHRU), London School of Hygiene and Tropical Medicine, Keppel Street, London WC1E 7HT, United Kingdom.*

Martin Williams, *Science Policy Unit, Defra, AQIP Programme, Area 3C Ergon House, 17 Smith Square, London SW1P 3JR, United Kingdom.*

Urban Air Pollution Climates throughout the World

OLE HERTEL* AND MICHAEL EVAN GOODSITE

ABSTRACT

The extent of the urban area, the local emission density, and the temporal pattern in the releases govern the local contribution to air pollution levels in urban environments. However, meteorological conditions also heavily affect the actual pollution levels as they govern the dispersion conditions as well as the transport in and out of the city area. The building obstacles play a crucial role in causing generally high pollutant levels in the urban environment, especially inside street canyons where the canyon vortex flow governs the pollution distribution. Of the pollutants dominating urban air pollution climates, particulate pollution in general together with gaseous and particulate polycyclic aromatic hydrocarbons (PAHs) and heavy metals are those where further field measurements, characterization and laboratory studies are urgently needed in order to fully assess the health impact on the urban population and provide the right basis for future urban air pollution management.

1 Introduction

In addition to other adverse health effects, air pollution is estimated to cause about 2 million premature deaths worldwide annually.[1] In this context particulate matter (PM) is generally believed to be the most hazardous of ambient pollutants, and it has been estimated that reducing ambient air concentrations of PM_{10} from 70 to $20 \, \mu g \, m^{-3}$ would lower the number of air quality related deaths by approximately 15%.[1] More than half of the world's population reside

Issues in Environmental Science and Technology, 28
Air Quality in Urban Environments
Edited by R.E. Hester and R.M. Harrison
© Royal Society of Chemistry 2009
Published by the Royal Society of Chemistry, www.rsc.org

in cities,[2] where the highest air pollution exposure[3] and associated negative health impact take place. Furthermore, the projections for the next 50 years indicate that the worldwide urban population will increase by two thirds.[2] Urban air pollution has been increasing in major cities, especially those found in developing countries (such as in: Brazil, Russia, India, Indonesia and China) as a result of rapid urbanisation. The cost to society of the associated health effects is significant and has been estimated to be approximately 2% of the Gross Domestic Product (GDP) in developed countries and 5% of GDP in developing countries (www.unep.org/urban_environment/issues/urban_air.asp). There may also be associated losses in productivity.[4]

1.1 Emission and Formation of Urban Air Pollution

Urban air pollution arises from the competition between emission processes which increase pollutant concentrations, and dispersion, advection and deposition processes that reduce and remove them. This chapter describes the differences in local urban pollutant levels between cities worldwide, and outlines how these differences in pollution levels reflect differences in emission densities and emission patterns, but also in pollutant dispersion and removal processes. The impact on pollution levels of the dispersion and removal processes are governed by the local meteorological conditions, which also vary heavily with the physical location of the city. Air pollution concentrations in an urban environment are naturally the result of local emissions as well as contributions from pollution transport from more remote sources (see Figure 1). The size of the city domain and the density of pollutant emissions govern the local contribution to urban air pollution.[5] Naturally, the temporal pattern in urban air pollution levels is a function of variations in the local releases, but just

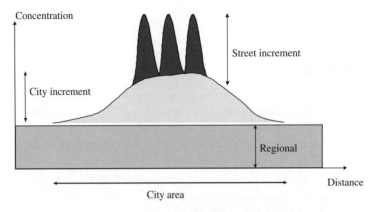

Figure 1 A schematic illustration of the air pollutant contribution from regional transport, the city area and the street traffic. The relative magnitude of the various contributions depends on the considered pollutant and the actual dispersion conditions (governed by the meteorology).

as important are the variations in the meteorological parameters that govern the dispersion and the pollutant transport in and out of the city.

Besides the influence from temporal variations in emissions and meteorology, the emission release height also plays an important role. Air pollution emitted from a high release height will in many cases be transported out of the urban area before being dispersed down to ground level; depending on the size of the urban domain. Urban industries, power plants and other sources for which the releases come from tall chimneys, contribute therefore only rarely to the local ground level air pollutant concentrations inside the urban area. These pollutant sources contribute primarily to the more regional air pollution.

Pollutant emissions related to vehicular transport, local domestic heating and smaller industries have low release heights [less than 10 m above ground level (a.g.l.)]. These releases are not diluted as efficiently as generally the case for emissions from tall release heights (more than 20 m a.g.l.). The contribution from "low" sources therefore often dominates the pollutant concentrations at ground level inside the urban area. A steady growth in vehicular transport and centralization of domestic heating have made road traffic the most important source of urban air pollution in many countries,[6] including most industrialized nations.

In respect to local contribution from different sectors there are generally significant differences between developed and developing countries. A comparison of two so-called mega-cities (Beijing and Paris) showed that aerosol particles and volatile organic compounds (VOCs) have a complex and multi-combustion source in Beijing, whereas a single traffic pollution source completely dominates the urban atmospheric environment in Paris.[7]

Indoor air quality is a major health concern. In the developing countries, emissions from household use of fossil fuels in the year 2000 was estimated to account for 1.6 million deaths, mainly among women and children in the poorest countries.[8]

In the present paper we focus on ambient air quality and related impact on human health. The actual ambient air pollutant load greatly varies from one city to another, but, generally, major urban areas throughout the world have poor air quality, and, among these, the cities in the developing countries face the greatest challenges. WHO has compiled a survey on typical ranges in ambient air concentrations of four indicator pollutants, which are summarized in Table 1.

1.2 Urban Pollution Levels and Indicators

The highest urban air concentrations of the classic pollutants like PM_{10} and SO_2 are found in Africa, Asia and Latin America, whereas the highest levels of secondary pollutants like O_3 and NO_2 are observed in Latin America and in some of the larger cities and urban air sheds in the developed countries. The environmental and human health impacts are particularly severe in cities of about 10 million or more inhabitants – also known as mega-cities.[9] Urban air

Table 1 Ranges in annual average urban ambient air concentrations ($\mu g\,m^{-3}$) of PM_{10}, NO_2, SO_2 and 1 hour average maximum concentrations of O_3 for different regions, based on a selection of urban data.[1]

	Annual average concentrations			1 h max concentration
Region	PM_{10}	NO_2	SO_2	O_3
Africa	40–150	35–65	10–100	120–300
Asia	35–220	20–75	6–65	100–250
Australia/New Zealand	28–127	11–28	3–17	120–310
Canada/United States	20–60	35–70	9–35	150–380
Europe	20–70	18–57	8–36	150–350
Latin America	30–129	30–82	40–70	200–600

pollution has become one of the main environmental concerns in Asia, and especially in China where the pollution load in mega-cities like Beijing, Shanghai, Guangzhou, Shenzhen and Hong Kong is substantial. In these cities, between 10 and 30% of days exceed the so-called Grade-II national air quality standards[10] by a factor of three to five times that of the WHO AQG (air quality grade). These cities have experienced a 10% growth in traffic each year over the last 5 to 6 years and, even with enhanced emission controls, NO_2 and CO concentrations have remained almost constant over the same period of time.

Use of air quality indices (AQIs) are common tools in environmental management. A description of widely used indices and how they are expressed mathematically is given in Gurjar *et al.*[11] AQIs can be designed to handle single or multi pollutants and may be used for comparing the loads in different cities or for describing the current load in relation to average loads or air quality standards and target values. In a multi component AQI (they applied the term MPI) ranging over mega-cities throughout the world, the highest MPI values were found for Dhaka, Beijing, Cairo and Karachi with values about double those of Delhi, Shanghai and Moscow[11] (Figure 2).

2   Sources in the Vicinity of the City

Airports are usually located in the vicinity of larger cities and often mentioned as potential sources of high pollution loads in the urban areas. In recent years, several studies have thus been carried out to determine the potential impact of airport emissions. These studies generally point at an influence from the road traffic going to and from the airport, whereas the impact of aircraft emissions has been found to be very limited. In a study carried out in Frankfurt Airport, signals from specific aircraft emissions generally could not be identified, whereas emissions from vehicle traffic on surrounding motorways had measurable impact on the air quality.[12] A study from Munich Airport had similar findings.[13] A study inside Heathrow Airport has shown that between 5 and

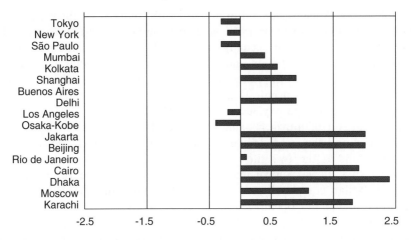

Figure 2   Mega-cities pollution indices (MPI) based on measurements of the classical
air pollutants and aggregated into an index for total pollution level (multi
pollutant). The plot is reproduced from Gurjar *et al.*[11]

30% of the local NO_x contribution is related to aircraft, whereas the remaining
95 to 70% is from road traffic.[14]

A recent study has indicated that ship traffic is responsible for about 60,000
lung cancer and cardiopulmonary deaths annually,[15] but this outcome is linked
to the contribution from ship emissions to the background PM load and not
particularly related to the urban air quality. Harbours may be a local source
contributing to urban pollution, but studies indicate that local road traffic often
dominates the contribution from harbours. A study in the harbour of Aberdeen
thus showed a gradient of increasing NO_2 and soot concentrations from the
harbour towards the city centre,[16] indicating the contribution from the harbour
had very limited impact on the local air quality in comparison with the emissions taking place in the urban environment.

Wood combustion in households is a growing concern in areas with many
wood stoves that have relatively high local emissions of PM in comparison
with other anthropogenic pollution sources. Investigations of wood combustion and air quality in developed countries like New Zealand,[17] Sweden,[18]
USA[19,20] and Denmark[21] have documented that residential wood combustion
may significantly elevate the local PM concentrations in outdoor air. As an
example, emission inventories for Denmark point at wood combustion as the
largest anthropogenic source of primary particle emissions.

3   Impact of the Geography, Topography and Meteorology

3.1   Geography

The location of the city has significant impact on the dispersion conditions,
mainly since it affects the local meteorological conditions. The classical

example is Los Angeles situated in a valley with frequent stagnant conditions during temperature inversions. The stagnant conditions lead generally to low wind speeds, and little air exchange between the valley and the surrounding areas. Hot and sunny climate and high emissions from traffic, industry and domestic heating makes the valley act like a large pollutant reaction chamber. This leads to high concentrations of photochemical products like ozone, nitrogen dioxide and peroxy acetyl nitrate (PAN).

A comparison of nitrogen oxide (NO_x) levels in the street Via Senato in Milan, Italy and the street Jagtvej in Copenhagen, Denmark showed similar concentrations at the two sites despite much higher traffic in the street of Copenhagen.[22] It was shown mainly to be a result of generally lower wind speeds in Milan compared with Copenhagen. High wind speeds and neutral conditions prevail in Copenhagen, whereas low wind speeds and stable or near stable conditions are frequent in Milan. Copenhagen has a cold coastal climate whereas Milan has a warm sub-tropical climate and the local wind conditions are furthermore affected by the location inside the Po Valley.

3.2 Topography

Some cities have characteristic wind systems as a result of local topography. An example of such effects is the rising air over a warm mountain side during daytime often leading to local formation of clouds and release of precipitation. During night the system turns around and the cooling of the air in the mountain valley leads to stable conditions that may cause local air pollution problems. The impact of Katabatic winds is another example, which affects cities along the Norwegian coast. The Katabatic winds are formed when cold air masses move down-slope (Katabatic is Greek for moving down hill) and meeting the colder snow and glacier covered areas, which then cool the air mass further, before the air floats out through a narrow cleft at the bottom of the hill. Usually the impact on air pollutant concentrations is moderate, but they may lead for example to high levels of local dust. Yet another example is the warm and dry Foehn wind formed on the back-side of a mountain chain, *e.g.* on the north side of the Alps. When the wind is forced over the mountain, the air is cooled and releases moisture. The air subsequently becomes warmer when it is moving down-hill again. This system may then form an inversion and, *e.g.* reduce dispersion of local air pollutants.

3.3 Meteorology

The ambient temperature in the urban atmosphere of larger cities is generally a couple of degrees Celsius higher than that found in the surrounding rural areas. This feature is termed the urban heat island effect,[23] and the explanation is that the city has a smaller albedo and therefore absorbs more energy compared with the surrounding rural areas. There is in addition a high consumption of energy inside the city, as a result of domestic heating and intense road traffic, which again contributes to release of heat. Finally, the buildings and other urban constructions form a shield for the wind, and this shielding leads to less cooling of the surfaces inside the city. Since the buildings act as heat reservoirs, the city has furthermore a less pronounced diurnal temperature variation compared with the rural area.

In calm weather, an urban circulation cell may be formed by warm air rising from the city. Some distance away, this heated air sinks and returns to the city at a low altitude. A similar phenomenon is known in coastal regions, where a sea breeze may be formed as a result of the temperature difference between the sea and land surfaces. A study in London showed that the heat island circulation over the city means that the wind speed is never below about $1\,\mathrm{m\,s^{-1}}$ (ref. 24). This study shows that the heat island effect is very important during low wind speed conditions in London where it may dominate the dispersion and thereby be limiting for the highest local air pollution concentrations, and this effect may thereby be the limiting factor for the highest pollution concentrations in the urban environment.

4 Pollutant Dispersion in Urban Streets (see also chapter by Salmond and McKendry)

Trafficked streets are air pollution hot spots in the urban environment (Figure 1). The concentration inside the urban street may be considered as the result of two contributions, one from emissions from the local traffic in the street itself and one from background pollution entering the street canyon from above roof level:[25]

$$c = c_b + c_s$$

where c is the concentration in the street, c_b the urban background contribution and c_s the contribution from traffic inside the street itself. The background contribution furthermore arises from two contributions; the first of these is the contribution from nearby sources in the urban area (typically this will mainly be traffic in surrounding streets), and the other contribution consists of regional (sources within a distance of a few hundred km) and long range transported (sources placed up to thousands of km away) pollution.

Naturally, the pollutant levels in the urban streets are strongly affected by traffic emissions taking place inside the street itself. However, the concentration level and the distribution of air pollution inside the street are to a large extent governed by the surrounding physical conditions. These physical conditions heavily affect the wind speed and especially the wind direction inside the street.[25] The special airflow generated inside the streets and around building obstacles may result in very different concentration levels at different locations in the street. The classical example is the street canyon vortex flow (Figure 3), which physically governs the pollutant distribution inside the street canyon. The street canyon is characterised by the presence of tall buildings on both sides

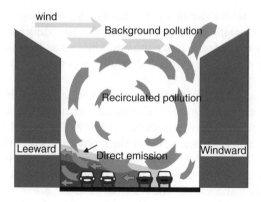

Figure 3   Illustration of the flow and dispersion inside a street canyon. In the situation
shown, the wind above roof level is blowing perpendicular to the street.
Inside the street canyon a vortex is created, and the wind direction at street
level is opposite to the wind direction above roof level. Pronounced differences (they may be up to a factor of 10) in air pollution concentrations on
the two pavements is the result of these flows.

of the street. Within the vortex flow relatively clean air from rooftop height is
drawn down at the windward face of the street, across the road at street level, in
the reverse of the wind direction at roof top, bringing pollutants in the road to
the leeward face of the canyon. This results in pollution concentrations up to 10
times higher on the leeward side compared with the windward side of the street.

5   Nitrogen Dioxide Pollution in Urban Areas (see also chapter by Bloss)

Residence time for an air packet in the vicinity of an urban street is usually of
the order of seconds to a few minutes,[25] depending on the street topography,
and therefore only very fast chemical conversions have time to take place.
For example the chemistry of nitrogen oxides [NO_x: the sum of nitrogen
monoxide (NO) and nitrogen dioxide (NO_2)] in urban streets may be described
by only two reactions: the reaction between ozone (O_3) and NO forming NO_2,
and the photo dissociation of NO_2:[26]

$$NO + O_3 \rightarrow NO_2 \tag{1}$$

$$NO_2 + h\nu \rightarrow NO + O(^3P) \tag{2}$$

$$O(^3P) + O_2 \rightarrow O_3 \tag{3}$$

$O(^3P)$ is ground state atomic oxygen. Reaction (3) is very fast, and for most
practical applications it may be disregarded. The products of reaction (2) may
thus be considered to be NO and O_3. NO_x is therefore mainly emitted as NO
and to a lesser extent, NO_2.

Long-term exposure to elevated NO_2 levels may decrease lung function and increase the risk of respiratory symptoms such as acute bronchitis, cough and phlegm, particularly in children,[27] whereas NO at current ambient air concentrations is considered to be harmless.

Previously, the fraction of NO_x directly emitted as NO_2 was only about 5 to 10% in countries with a small fraction of diesel engines. Due to the use of catalytic converters and an increasing number of diesel engines with high fraction of NO_2 in the exhaust, this value may in some regions be up to as much as 40%.[28] This extremely simple chemical mechanism of two reactions and a direct emission describes very well the concentrations of NO_2 inside urban streets,[26] and *e.g.* for Northern European cities it may also be applied for describing NO_2 concentrations in urban background air.[5]

The use of catalytic converters has in recent years led to a significant reduction in NO_x concentrations in urban streets of many industrialised countries. However, for several reasons NO_2 levels have not followed the same trends (Figure 4). Part of the explanation is the chemical conversion of NO to NO_2 in the reaction with O_3, but another explanation is an increased fraction of NO_2 in the NO_x emission from vehicles with catalytic converters. Despite the overall reduction in NO_x emissions this is contributing to elevated NO_2 concentrations. Prognoses for the development in NO_2 concentrations in Denmark indicate that the future exhaust standards for road traffic vehicles will solve the current problems of complying with EU limit values. It is therefore important that similar emission restrictions take place in developing countries in the future in order to solve the problem of exposure of the population to elevated NO_2 levels here also.

6 Particle Pollution in Urban Areas

Ambient urban air contains a complex mixture of particles of varying sizes and chemical composition.[29–33] The size is crucial for the atmospheric fate[34] as well as the human health impact, since it governs the particles' atmospheric behaviour as well as their deposition in the human respiratory system.[35] Particles in ambient air typically appear in three rather distinct size classes (or modes) usually termed ultrafine (diameter: 0.01–0.1 µm or 10–100 nm), fine (diameter: 0.1–2.5 µm) and coarse (diameter: > 2.5 µm) mode particles (Figure 5).

6.1 Particle Mass Concentrations

Only the mass concentration of particles < 10 µm in aerodynamic diameter (PM_{10}) is generally used as an indicator for suspended particulate matter (*e.g.* regulated in EU directives) and routinely measured at many locations throughout the world. Particle mass is dominated by particles > 0.1 µm in diameter. The particles that appear in traffic exhaust are found mainly in the ultrafine fraction and include elemental carbon (EC) as well as organic carbon (OC).[36] These particles contribute considerably to the PM number

Ole Hertel and Michael Evan Goodsite

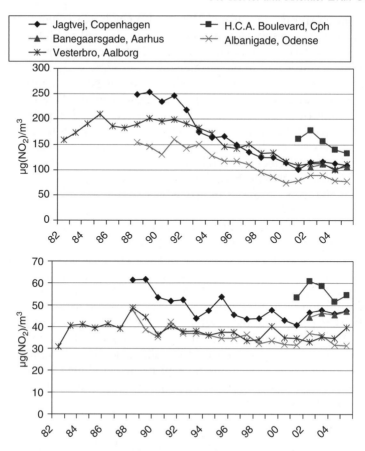

Figure 4   The measured trend in annual mean concentrations of NO_x and NO_2 [both shown in $\mu g(NO_2)/m^3$] at the street stations in the largest Danish cities: Copenhagen, Aarhus, Odense and Aalborg. Upper plot shows NO_x and lower plot the NO_2. The plots include measurements from the time period 1982 to 2005, and illustrate the decrease in NO concentrations that results from increasing number of vehicles with catalytic converters in the Danish car parks, but also that this decrease is not reflected in the NO_2 concentrations that have remained more or less constant during this time period.[63]

concentrations but only little to the PM mass. There are health studies that show relationships between both acute and long-term health effects and ultrafine particles (particle sizes $<0.1\,\mu m$),[37] but other studies are less conclusive concerning health effects of ultrafine particles (expressed as particle number concentrations). Although ultrafine particles give a minor contribution to mass concentrations, they represent most of the particles in terms of number concentration. Most studies of health effects of ambient air particles have been related to particle mass, but even here the mechanisms are not fully understood and the need for more studies has been underlined.[38]

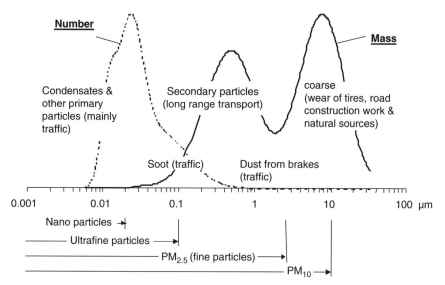

Figure 5 The typical size distribution of particles in urban air given in both mass and number concentration. The horizontal axis is the particle diameter in μm. The full line is mass distribution, dominated by the coarse and secondary particles. The dashed line is the number distribution, dominated by ultrafine particles. Note that one particle with a diameter of 10 μm has the same weight as 1 billion particles with a diameter of 0.01 μm.[64]

In busy streets, a significant fraction of the particle pollution originates from traffic.[39] The direct emission from car exhaust contains particles formed inside the engine as well as in the air just after the exhaust pipe. The latter depends on the sulfur content in the fuel; studies have shown that reducing the sulfur content in diesel significantly reduces the particle number concentrations in urban streets. The directly emitted particles are found mainly in the ultrafine particle fraction. However, traffic also contributes to mechanically formed particles in the fine and especially the coarse fraction. The particles in the coarse fraction are produced from wear of tyres and road surface material as well as re-suspended dust. The particles from the brakes contribute similar amounts to the fine and the coarse fraction.

For particle mass (PM$_{10}$), long-range transport is usually the dominating source for regional background levels. Danish studies have shown that less than 10% of the urban PM$_{10}$ originate from local urban sources.[39] For particle numbers as well as NO$_x$ a much larger difference between rural, urban and curb side levels is observed, indicating a large contribution from local traffic sources (Figure 6).

6.2 Particle Number Concentrations

A striking feature of urban particles is the often very high correlation between concentrations of NO$_x$ and total particle number, indicating that both

Ole Hertel and Michael Evan Goodsite

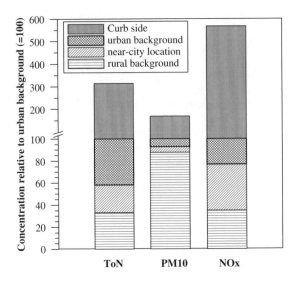

Figure 6   Comparison of average concentrations of total particle number (ToN), particle mass (PM_{10}) and NO_x at rural, near-city, urban and curb side stations relative to urban background levels in the Copenhagen area. The concentration bars are stacked so that only the additional contributions are marked with the pattern shown in the legend. Note that the scale of the vertical axis changes at 100. Adapted from Ketzel *et al.*[39]

compounds originate from the same (traffic) source (one study[39] found R > 0.9). They are emitted in a similar ratio (particle number : NO_x) from the different traffic categories, *i.e.* high NO_x emitters (diesel vehicles, especially heavy-duty vehicles) are also high particle emitters (when these are expressed in particle number concentrations).[40] Model calculations such as those with the Danish Operational Street Pollution Model (OSPM)[41] have been shown to reproduce well the observed particle number, when treating particles as inert tracers (disregarding transformation and loss processes).[42] The particle emission factors depend on ambient temperature with higher emissions at lower temperatures, which is accounted for in the simulations.[43]

Long-range transport contributes significantly to fine fraction particles and leads to the main part of particulate sulfate and ammonium and a large part of particulate nitrate. These secondary particles are formed from anthropogenic sulfur dioxide (SO_2), ammonia (NH_3) and NO_x emissions,[44] and often constitute more than 30% of the PM_{10}. Another part of the particulate nitrate appears in the coarse fraction, which also contains contributions from sea spray and re-suspended dust (including road dust)[45,46] that has a relatively large mass and quickly deposits by gravitational settling. Coarse particles, therefore, have a short lifetime in the atmosphere compared with fine particles. Combustion in wood stoves is a source of particle pollution, which contributes about 90% of the total particle emissions attributed to domestic heating in Denmark. As an example, road traffic and use of wood stoves are the largest Danish sources of particle exposure of the population, due to the low release height and because the emissions take place where people live. The particles emitted from wood stove combustion are soot particles with high contents of polycyclic aromatic hydrocarbons (PAH).

6.3 Importance of Measurement Location

People in temperate climates spend a significant part of their time indoors. Exposure to air pollution in the home is thus an important fraction of their overall exposure. A Danish study in an uninhabited apartment in central Copenhagen revealed that particle pollution inside the apartment was to some extent linked to the activity level of the neighbouring apartments.[47] This indicates that sources in the neighbouring environment must also be considered in the analysis of overall exposure.

When comparing and analysing observed levels of PM in different cities one must take into account that the location of the monitoring stations may be very different, and there is a risk of comparing sites in the vicinity of large pollution sources with sites at some distance from local sources.

6.4 Contribution from Natural Processes

Data collected in a WHO study (Figure 7), indicate that PM_{10} concentrations in Asia and Latin America are higher than are observed in Europe and North America.[1] The highest particle levels are observed in Asia and are attributed to forest fires, poor fuel quality and aeolian (windblown) dust. Wind erosion originating especially in the deserts of Mongolia and China contributes to the general level of PM in the region.

7 Polycyclic Aromatic Hydrocarbons (PAH) in Urban Areas

7.1 Sources and Emissions

Polycyclic aromatic hydrocarbons are a group of chemicals that are formed during incomplete burning of coal, oil, gas, wood, garbage, or other organic substances, such as tobacco and charbroiled meat. There are more than 100 different PAHs. PAHs generally occur as complex mixtures (*i.e.* as part of combustion products such as soot), not as single compounds. They usually occur adventitiously, but they can be manufactured as individual compounds for research purposes; however, not as the mixture found in combustion products. They can also be found in materials such as crude oil, coal, coal tar pitch, creosote, and roofing tar. A few PAHs are used in medicines and in the production of dyes, plastics, and pesticides. Others are contained in asphalt used in road construction.

Given the numerous sources of emissions of combustion products in urban (especially in developing countries) areas, PAHs are generally considered

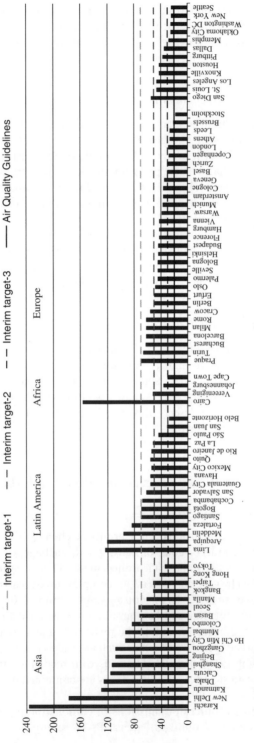

Figure 7 PM$_{10}$ concentrations (µg m^{-3}) in selected large cities throughout the world. The plot has been derived on basis of WHO (2006).[1] The WHO Air Quality Guideline (AQG) [shown in the figure] defined at the lowest level at which total, cardiopulmonary and lung cancer mortality has been shown to increase with more than 95% confidence in response to long-term exposure. Shown also are the WHO interim target values 1 to 3.

ubiquitous in the atmospheric environment and the urban environment in general. They occur in air, either attached to dust particles or as solids in air-borne soil or sediment. PAHs are also a common product of combustion from common sources such as motor vehicles, and other gas-burning engines, wood-burning stoves and furnaces, cigarette smoke, industrial smoke or soot, and charcoal-broiled foods.[48,49] This is of great concern due to the mutagenic[50] and carcinogenic[51] properties of PAHs. PAHs of three or more benzene rings have a low vapour pressure and low solubility in water. Therefore they are present in ambient air both as gaseous compounds and as material adsorbed on particles. Lighter PAHs are accordingly almost always observed in gas phase, whereas heavier PAHs are generally observed on particles. The United States Agency for Toxic Substances (ATSDR, 1995) has listed 17 PAHs as of priority concern with respect to their toxicological profile. PAHs have been studied in urban and other areas and appear to represent a fraction of a percent of the ambient particle mass.[52] However, there are compelling arguments that the previously commonly employed measurement techniques produced artefacts and despite their toxicological profile, no country has mandatory guidelines with respect to ambient air quality standards.[53] The reader is referred to a recent review paper for greater detail and the current state of the art.[53]

Commercial production has been found not to be a significant source of PAHs in the environment.[54] The primary sources of many PAHs in ambient air is the incomplete combustion of wood and other fuels.[55] Natural sources include volcanoes, forest fires, crude oil, and shale oil. Only three of the 7 PAHs included in the ATSDR profiles are produced commercially in the United States in quantities greater than research level: acenaphthene, acenaphthylene, and anthracene.[49] Studies should be conducted to see if commercial production (accounting for a products' total life cycle) is a significant source of PAHs to the urban environment.

7.2 Sampling Artefacts

In order to reduce the risk of sampling artefacts,[52] in the spring of 2003 the Mexico City Metropolitan Area (MCMA) campaign employed three inde-pendent methods to measure particle-bound PAHs.[56,57] They found peak concentrations of PAHs on the order of $120 \, \text{ng m}^{-3}$ during the morning rush hour. Accordingly in urban areas, depending on the fuel sources of the vehicles and the use of catalytic converters, motor vehicle traffic is a significant source of PAHs with the balance of the remainder from trash and biomass burning. As an example, one specific study found that 20% of the vehicles were accounting for 50% of the PAH emissions.[58] Like other atmospheric gaseous and particulate constituents, PAHs are removed from the atmosphere by wet or dry deposition and may also be converted or degraded in heterogeneous pro-cesses. A very rapid decay of surface PAHs in the morning photochemistry has been reported.[52] However, in this study, it could not be ruled out that sur-factants coating the particles may have affected the sensitivity of the applied instrument. This is another example of the complexity of PAH measurements.

There may be significant concentrations of very toxic and very reactive PAHs in the MCMA atmosphere that are missed due to filter reaction artefacts.[57]

7.3 Long Range Transport

Long range transport of PAHs are a concern. PAHs are designated as one of the persistent toxic substances in central and northeast Asia under the Stockholm Convention (UNEP, 2002). It has been demonstrated that transformation processes can lead to PAHs which are more toxic than their precursors.[56] Processes describing the transformation and fate of PAHs in the urban environment are needed to further quantify the magnitude of the health risks associated with this toxic pollutant source.

One study has analysed the source apportionment of particulate PAHs at Seoul, Korea, during a measurement campaign between August 2002 and December 2003,[59] by applying the US EPA (2004) chemical mass balance (CMB) model; as Seoul is in the atmospheric footprint of major coal burning industries and power plants of both China and Japan, as well as contributing itself to the Northeast Asia particle footprint. In Seoul, similar to MCMA, gasoline and diesel vehicles accounted for 31% of the measured PAHs, *i.e.* the major sources. Daily and seasonal variations were noted and attributed to differences in biomass burning and coal (heating) with a 19% difference in the total concentrations observed between fall (63%) and winter (82%). The sources had an inverse profile away from the city, which the authors attribute via source analysis to long range transport of atmospheric pollutant (LRTAP) PAHs and their precursors from sources in China or North Korea.

7.4 Future Requirements

The Korean study, as well as the other studies discussed above, document the urgent need for better understanding of PAHs in the urban atmosphere as well as their transformation, transport and conversion. Although air quality of mega-cities and urban areas fall within the jurisdiction of local governments, the studies provide compelling evidence that efforts at international levels to regulate LRTAP must specifically look at PAHs and their precursors. This will require an international, strategic joint effort amongst Asia, Europe and North America to investigate these complicated mechanisms, and analyse the data. There are often many air quality monitoring stations (198 in Korea) in countries collecting data, but for PAHs, they are not well established, and levels are reported in only a few studies, and these are usually from campaigns that predate the publications by years, owing to the complexity of analysing the data.

8 Trace Elements, including Heavy Metals in Urban Areas

Trace elements including heavy metals (HM) are ubiquitous in the atmosphere of urban areas and many are classified as Hazardous Air Pollutants (HAP).

They represent serious environmental and health risks and are especially of concern in industrial and urban climates, with mega-cities in developing nations being the obvious sites of concern for populations at risk for trace element exposure. One study has provided trends and levels of trace metals in three Danish cities as well as at background sites.[60] The levels reported are less than literature values for mega-cities, as expected, but in many cases the trends are similar. More studies need to be accomplished with respect to trends in developing and mega-cities.

8.1   Heavy Metals

Western nations have tightened and continue to restrict emissions of such trace elements as mercury, but despite decades of history of efforts to regulate trace metals they still represent a real and pressing health and environmental hazard in the US (US EPA, 2006) and certainly in other western nations and other countries as well. Humans are exposed to metals via ambient air inhalation, and consuming contaminated food or water, as well as, in the case of lead and children, chewing on lead painted toys, and exposure from walls or furniture.

Several metals aside from mercury and lead are classified by the United States Clean Air Act as HAPs: chromium, manganese, nickel, and cadmium. The United States Environmental Protection Agency (US EPA) lists many trace metals as among the worst urban air toxics (www.epa.gov/ttn/atw/nata/34poll.html, 28 November, 2008).

8.2   Trace Elements

Trace elements and metals were of great concern in western cities prior to lead being banned as an anti-knock agent in gasoline in the mid 70s, as well as other regulations regarding the use of lead in, for example, house paints. The observed high levels and effects have led to extensive monitoring with con-sequential regulations and international agreements that have greatly reduced the concentrations.

The sources of emission into the urban atmosphere of commonly measured trace elements: Be, Co, Hg, Mo, Ni, Sb, Se, Sn, and V, with smaller con-centrations of As, Cr, Cu, Mn, and Zn, are primarily human activities, espe-cially the combustion of fossil fuels and biomass, industrial processes and waste incineration.[61]

Certain trace elements have significant natural sources such as: from sea spray near coastlines, dust from aeolian processes and weathering, especially noted in Asia, and volcanoes with local, regional and global effects. On any given day however, the main source of trace elements in urban environments will be through anthropogenic activities, and national regulations, especially with regard to limits in gasoline, and emissions from coal fired power plants, as well as industrial, especially metallurgic and (in the case of mercury) chlor-alkali processes, mean that developing nations have higher levels in general than western nations in their urban environments.

Trace metals have been measured in nearly all aerosol size fractions. It is therefore of paramount importance to characterize the particle size distribution and relate these to the potential adverse health effects in the urban population.[61]

As with PAHs, there are seasonal variations of trace metals in PM_{10} and $PM_{2.5}$ observed in campaigns conducted seasonally, with higher winter values of nearly all trace elements suggesting a significant source of particles from domestic heating in most temperate urban areas, and/or less efficient dispersion of emissions in winter.

8.3   Recommendations for Modelling

A regional model for atmospheric photochemistry and particulate matter was applied to predict the fate and transport of five trace metals; lead, manganese, total chromium, nickel, and cadmium, over the continental United States during January and July 2001.[62] This study may be used to summarize the state of the art of the modelling of trace metals and the limitations of trace metal inventories. The authors recommend research in order to improve the model results on emission data for aerial suspension of particles and biomass burning. They note that including these sources will require models of aerial suspension and combustion as well as composition data for fuel and soil over the modelled location. Their recommendations will enable better modelling of especially lead and manganese. They recommend further research to better quantify anthropogenic emissions of chromium, nickel and cadmium [in the US National Emission Inventory (NEI)]. It is our opinion that their recommendations are valid for most NEI.

9   Conclusions

Urban air pollution is a complex and dynamic mixture of gaseous and particulate pollutants with both daily and seasonal variation due to both anthropogenic activity levels and weather conditions. The highest urban air pollutant concentrations of some of the most studied pollutants like PM_{10} and SO_2 are found is Africa, Asia and Latin America. The highest levels of photochemical pollutants like O_3 and NO_2 are observed in Latin America and some of the developed countries. The negative health effects of urban air pollution are well documented and found to be particularly severe in the mega-cities, where quality of life is lessened due to air pollution and the possibility of reduced productivity from toxics should be investigated.[4] The actual pollution load in a given urban area is the result of both local emissions and transport from both nearby and more remote sources. The location of the city is very important for the local dispersion conditions, which are governed by meteorology but are also heavily affected by topographical conditions (*e.g.* effects of coastline, mountains, valleys *etc.*). Studies of the impact of local harbours and airports have generally pointed at a limited influence on urban air quality, and that the impact is mainly related to the road traffic that these facilities are generating.

Although the health effects of urban air pollution have been documented in numerous studies, there are still major unknowns in this regard. This review points at an urgent need for field studies dedicated to a better characterisation of urban particle pollution together with studies focussing at gaseous and particulate PAHs and trace elements such as heavy metals. Such studies are needed for the full assessment of the health impact on the urban population and for providing the necessary basis for future urban air pollution management.

In this chapter we have not addressed the synergistic impact on human health of the chemical cocktail arising from diverse sources in the various micro-environments where the population reside in daily life. The negative health effects of air pollution may be enhanced by exposure to environmental tobacco smoking, indoor sources like cooking, candles, stoves *etc.* but also by exposures to non-airborne agents in textiles, food *etc.* For this complex interaction of different exposures and their impact on human health, the reader is referred to the literature.

References

1. *WHO Air Quality Guidelines for Particulate Matter, Ozone, Nitrogen Dioxide and Sulfur Dioxide – Summary of Risk Assessment*, WHO, 2006.
2. *UNFPA State of World Population 2007 – Unleasing the Potential of Urban Growth*, UNFPA, 2007.
3. O. Hertel, F. A. A. M. de Leeuw, O. Raaschou-Nielsen, S. S. Jensen, D. Gee, O. Herbarth, S. Pryor, F. Palmgren and E. Olsen, *Pure Appl. Chem.*, 2001, **73**, 933–958.
4. L. D. Hylander and M. E. Goodsite, *Sci. Total Environ.*, 2006, **368**, 352–370.
5. R. Berkowicz, F. Kunzli, R. Kaiser, S. Medina, M. Studnika, O. Chanel, P. Filliger, M. Herry, F. Horak, V. Puybonnieux-Texier, P. Quenel, J. Schneider, R. Reethaler, J. C. Vergnaud and H. Sommer, *Environ. Monit. Assess.*, 2000, **65**, 259–267.
6. N. Kunzli, *et al., Lancet*, 2000, **356**, 795–801.
7. V. Gros, J. Sciare and T. Yu, *C. R. Geosci.*, 2007, **339**, 764–774.
8. K. K. Smith, *et al., Indoor Air Pollution from Household Use of Fossil Fuels*, WHO, 2004.
9. B. R. Gurjar and J. Lelieveld, *Atmos. Environ.*, 2005, **39**, 391–393.
10. C. K. Chan and X. Yao, *Atmos. Environ.*, 2008, **42**, 1–42.
11. B. R. Gurjar, T. M. Butler, M. G. Lawrence and J. Lelieveld, *Atmos. Environ.*, 2008, **42**, 1593–1606.
12. H. J. Crecelius and M. Sommerfeld, *Gefahrstoffe Reinhalt. Luft*, 2005, **65**, 49–54.
13. P. Suppan and J. Graf, *Int. J. Environ. Poll.*, 2000, **14**, 375–381.
14. F. Farias and H. ApSimon, *Environ. Model. Software*, 2006, **21**, 477–485.
15. J. J. Corbett, *et al., Environ. Sci. Technol.*, 2007, **41**, 8512–8518.

16. I. L. Marr, D. P. Rosser and C. A. Meneses, *Atmos. Environ.*, 2007, **41**, 6379–6395.

17. H. B. Wang and D. Shooter, *Atmos. Environ.*, 2002, **36**, 3519–3529.

18. E. Hedberg and C. Johansson, *J. Air Waste Manag. Assoc.*, 2006, **56**, 1669–1678.

19. K. Sexton, K. S. Liu, S. B. Hayward and J. D. Spengler, *Atmos. Environ.*, 1985, **19**, 1225–1236.

20. N. N. Maykut, J. Lewtas, E. Kim and T. V. Larson, *Environ. Sci. Technol.*, 2003, **37**, 5135–5142.

21. M. Glasius, M. Ketzel, P. Wåhlin, R. Bossi, J. Stubkjær, O. Hertel and F. Palmgren, *Atmos. Environ.*, 2008, **42**, 8686–8697.

22. E. Vignati, R. Berkowicz and O. Hertel, *Sci. Total Environ.*, 1996, **190**, 467–473.

23. I. Eliasson and B. Holmer, *Theo. App. Climatol.*, 1990, **42**, 187–196.

24. M. Rigby and R. Toumi, *Atmos. Environm.*, 2008, **42**, 4932–4947.

25. R. Berkowicz, F. Palmgren, O. Hertel and E. Vignati, *Sci. Total Environ.*, 1996, **190**, 259–265.

26. F. Palmgren, R. Berkowicz, O. Hertel and E. Vignati, *Sci. Total Environ.*, 1996, **190**, 409–415.

27. M. Brauer, G. Hoek, P. van Vliet, K. Meliefste, P. H. Fischer, A. Wijga, L. P. Koopman, H. J. Neijens, J. Gerritsen, M. Kerkhof, J. Heinrich, T. Bellander and B. Krunekreef, *Am. J. Respir. Critical Care Med.*, 2002, **166**, 1092–1098.

28. D. C. Carslaw and S. D. Beevers, *Atmos. Environ.*, 2005, **39**, 167–177.

29. D. C. S. Beddows, M. C. S. Beddows, R. J. Donovan, R. M. Harrison, R. P. Kinnersley, M. D. King, D. H. Nicholson and K. C. Thompson, *J. Environ. Monitor.*, 2004, **6**, 124–133.

30. R. M. Harrison, A. M. Jones and R. G. Lawrence, *Atmos. Environ.*, 2004, **38**, 4531–4538.

31. P. Wåhlin, R. Berkowicz and F. Palmgren, *Atmos. Environ.*, 2006, **40**, 2151–2159.

32. D. Salcedo, T. B. Onasch, K. Dzepina, M. R. Canagaratna, Q. Zhang, J. A. Huffman, P. F. DeCarlo, J. T. Jayne, P. Mortimer, D. R. Worsnop, C. E. Kolb, K. S. Johnson, B. Zuberi, L. C. Marr, R. Volkamer, L. T. Molina, M. J. Molina, B. Cardenas, R. M. Bernabe, C. Marques, J. S. Caffney, N. A. Marley, A. Laskin, V. Shutthanandan, Y. Xie, W. Brune, R. Lsher, T. Shirley and J. L. Jimenez, *Atmos. Chem. Phys.*, 2006, **6**, 925–946.

33. M. Viana, T. A. J. Kuhlbusch, X. Querol, A. Alastuey, R. M. Harrison, P. K. Kopke, W. Winiwarter, A. Vallius, S. Szidat, A. S. H. Prevot, C. Heuglin, H. Bloemen, P. Wåhlin, R. Vecchi, A. I. Miranda, A. Kasper-Giebl, W. Maenhaut and R. Hitzenbergerq, *J. Aerosol Sci.*, 2008, **39**, 827–849.

34. R. M. Harrison, J. P. Shi, S. H. Xi, A. Khan, D. Mark, R. Kinnersley and J. X. Yin, *Philos. Trans. R. Soc. London, Ser. A: Math. Phys. Eng. Sci.*, 2000, **358**, 2567–2579.

35. J. Løndahl, J. Pagels, E. Swietlicki, J. C. Zhou, M. Ketzel, A. Massling and M. Bohgard, *J. Aerosol Sc.*, 2006, **37**, 1152–1163.

36. R. M. Harrison and J. X. Yin, *Atmos. Environ.*, 2008, **42**, 1413–1423.
37. A. Peters and H. E. Wichmann, *Epidemiol.*, 2001, **12**, S97.
38. R. M. Harrison and J. X. Yin, *Sci. Total Environ.*, 2000, **249**, 85–101.
39. M. Ketzel, P. Wåhlin, A. Kristensson, E. Swietlicki, R. Berkowicz, O. J. Nielsen and F. Palmgren, *Atmos. Chem. Phys.*, 2004, **4**, 281–292.
40. R. M. Harrison, R. Tilling, M. S. C. Romero, S. Harrad and K. Jarvis, *Atmos. Environ.*, 2003, **37**, 2391–2402.
41. R. Berkowicz, *Environ. Monitor. Assess.*, 2000, **65**, 323–331.
42. M. Ketzel and R. Berkowicz, *Atmos. Environ.*, 2004, **38**, 2639–2652.
43. M. Ketzel, G. Omstedt, C. Johansson, I. Düring, M. Pohjola, D. Öttl, L. Gidhagen, P. Wåhlin, A. Lohmeyer, M. Haakana and R. Berkowicz, *Atmos. Environ.*, 2007, **41**, 9370–9385.
44. B. H. Baek and V. P. Aneja, *J. Air Waste Manag. Assoc.*, 2004, **54**, 623–633.
45. T. A. Pakkanen, V. M. Kerminen, R. E. Hillamo, M. Makinen, T. Makela and A. Virkkula, *J. Atmos. Chem.*, 1996, **24**, 189–205.
46. T. A. Pakkanen, *Atmos. Environ.*, 1996, **30**, 2475–2482.
47. T. Schneider, K. A. Jensen, P. A. Clausen, A. Afshari, L. Gunnarsen, P. Wåhlin, M. Glasius, F. Palmgren, O. J. Nielsen and C. L. Fogh, *Atmos. Environ.*, 2004, **38**, 6349–6359.
48. *IARC Monographs on the Evaluation of the Carcinogenic Risk of Chemicals to Humans*, WHO, International Agency for Research in Cancer, 1983.
49. *ATSDR Toxicological Profile for Polycyclic Aromatic Hydrocarbons (PAHs)*, Dep. Health Human Serv., Pub. Health Serv., 1995.
50. I. C. Hanigan, F. H. Johnston and G. G. Morgan, *Environ. Health*, 2008, **7**, 42.
51. M. F. Denissenko, A. Pao, M. S. Tang and G. P. Pfeifer, *Science*, 1996, **274**, 430–432.
52. L. T. Molina, C. E. Kolb, B. de Foy, B. K. Lamb, W. H. Brune, J. L. Jimenez, R. Ramos-Villegas, J. Sarmiento, V. H. Paramo-Figueroa, B. Cardenas, V. Gutierrez-Avedoy and M. J. Molina, *Atmos. Chem. Phys.*, 2007, **7**, 2447–2473.
53. K. Ravindra, R. Sokhi and R. Van Grieken, *Atmos. Environ.*, 2008, **42**, 2895–2921.
54. J. Perwak, M. Byrne, S. Coons, M. Goyer, J. Harris, P. Cruse, R. Derosier, K. Moss and S. Wendt, *Exposure and Risk Assessment for Benzo(a)pyrene and other Polycyclic Aromatic Hydrocarbons. 4*, EPA, Off. Water Reg. and Stds., 1982.
55. *HSDB Hazardous Substances Data Bank*, National Library of Medicine, National Toxicology Program (via TOXNET), 1994.
56. L. C. Marr, K. Dzepina, J. L. Jimenez, F. Reisen, H. L. Bethel, J. Arey, J. S. Gaffney, N. A. Marley, L. T. Molina and M. J. Molina, *Atmos. Chem. Phys.*, 2006, **6**, 1733–1745.
57. K. Dzepina, J. Arey, L. C. Marr, D. R. Worsnop, D. Salcedo, Q. Zhang, T. B. Onasch, L. T. Molina, M. J. Molina and J. L. Jimenez, *Int. J. Mass Spectrom.*, 2007, **263**, 152–170.

58. C. Jang and J. Lee, *Proc. Inst. Mech. Eng., Part D: J. Automobile Eng.*, 2005, **219**, 825–831.
59. J. Y. Lee and Y. P. Kim, *Atmos. Chem. Phys.*, 2007, **7**, 3587–3596.
60. K. Kemp, *Nucl. Instrum. Methods Phys. Res., Section B: Beam Interact. Mater. Atoms*, 2002, **189**, 227–232.
61. S. Rajsic, Z. Mijic, M. Tasic, M. Radenkovic and J. Joksic, *Environ. Chem. Lett.*, 2008, **6**, 95–100.
62. W. T. Hutzell and D. J. Luecken, *Sci. Total Environ.*, 2008, **396**, 164–179.
63. O. Hertel, T. Ellermann, F. Palmgren, R. Berkowicz, P. Løfstrøm, L. M. Frohn, C. Geels, C. A. Skjøth, J. Brandt, J. Christensen, K. Kemp and M. Ketzel, *Environ. Chem.* 2007, **4**, 65–74.
64. O. Hertel, S. S. Jensen, M. Hvidberg, M. Ketzel, R. Berkowicz, F. Palmgren, P. Wåhlin, M. Glasius, S. Loft, P. Vinzents, O. Raaschou-Nielsen, M. Sørensen and H. Bak, in *Assessing the Impact of Traffic Air Pollution on Human Exposures and Linking Exposures to Health Effects Traffic, Road Pricing and the Environment*, ed. C. Jensen-Butler, B. Madsen, O.A. Nielsen and B. Sloth, Springer Press, 2007.

Influences of Meteorology on Air Pollution Concentrations and Processes in Urban Areas

JENNIFER A. SALMOND AND I. G. McKENDRY

ABSTRACT

In urban areas pollutant concentrations are determined by the balance between pollutant emissions, and the state of the atmosphere which determines pollutant transport and dispersion processes. In this chapter we demonstrate that although the nature and characteristics of urban emissions can be variable in time and space, it is changes in pollutant dispersion pathways both locally and regionally that often determine the temporal and spatial patterns of atmospheric composition in urban areas.

Due to the complexity of the urban atmosphere it is useful to look at the relations between urban meteorology and air pollution at a variety of different scales. At scales ranging from individual buildings and street canyons to the entire city, microscale mechanical and thermally driven turbulence dominates local dispersion processes. However, these processes operate within an hierarchy of larger scales which provide the background state of the atmosphere that modulates air quality within urban areas. Day-to-day variations in urban air quality are determined in large part by processes operating at scales greater than the urban environment. These meteorological processes extend from the regional scale flows such as sea breezes, to the synoptic scale (cyclonic and anticyclone systems that give our day-to-day weather), and finally to the hemispheric and global scales which permit long-range transport of pollutants.

Generally, as the ability of the urban atmosphere to disperse pollutants horizontally and vertically increases, pollutant concentrations decrease.

Issues in Environmental Science and Technology, 28
Air Quality in Urban Environments
Edited by R.E. Hester and R.M. Harrison
© Royal Society of Chemistry 2009
Published by the Royal Society of Chemistry, www.rsc.org

However, complex feedback loops exist between pollutant emission rates, urban form and meteorological processes, local and regional pollutant transport rates and atmospheric chemistry such that individual parameters are rarely independent of each other. Thus an understanding of local, urban and regional scale atmospheric processes is fundamental to a comprehensive evaluation of air pollution in urban areas.

1 Introduction

Urban air quality is inherently variable in time and space. Concentrations of air pollutants are primarily determined by the balance between emissions rates and the ability of the atmosphere to transport and disperse pollutants. In urban areas the high density of economic activities leads to intensive resource consumption and the release of large quantities of pollutants at a local scale. Whilst the immediate impact of these pollutants is undoubtedly greatest at local scales, urban plumes may extend for many hundreds of kilometres downwind, thus their contribution to pollution at regional and global scales cannot be ignored. As the world population becomes increasingly urbanised there is a need to improve our understanding of the processes influencing urban air quality in order to minimise population exposure and to develop coherent air pollution abatement strategies.

There is considerable geographical variation in ambient pollution concentrations in urban areas around the world (Figure 1). City size is a poor indicator of high levels of air pollution at a global scale due to variations in the

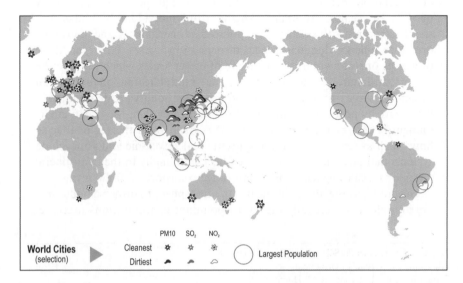

Figure 1   Geographical variation in urban air quality from selected cities around the world. Based on survey data provided by the World Bank (2007).[1]

availability and application of pollutant abatement technology, industrialisation and level of economic development. Based on annual mean measurements of three pollutants [sulfur dioxide (SO_2), particulate matter smaller than 10 microns in diameter (PM_{10}), and nitrogen dioxide (NO_2)] from a single monitoring point within each city it is possible to compare urban air quality around the world.[1] Five cities appear in the top twenty polluted cities for all three pollutants. These cities are all in China (Beijing, Lanzhou, Anshan, Chengdu and Shenyang) and reflect the dominance of both local and regional sources of industrial, domestic and vehicle emissions. The success of pollutant abatement strategies designed to limit domestic and industrial emissions within urban areas in North American and Europe is reflected in the absence of these cities in the SO_2 and PM_{10} categories. The dominant source of pollutants in many of these cities today is vehicle emissions.[2] Local and regional transport of vehicle emissions account for the high concentrations of NO_2 observed in Milan, Sydney, New York, London and Los Angeles. Three cities are among the top twenty cleanest cities for all three pollutants: Perth (Australia), Stockholm (Sweden) and Auckland (New Zealand). These are all cities with smaller populations, little industry and are located away from dominant regional sources of pollution.

Although the nature and characteristics of urban emissions can be variable in time and space, it is changes in pollutant dispersion pathways both locally and regionally that often determine the temporal and spatial patterns of atmospheric composition in urban areas.[3] Periods of poor air quality in urban areas are associated with meteorological conditions which limit dispersion horizontally (due to low wind speeds) and vertically.[4] The ability of the atmosphere to mix pollutants vertically is determined by the *stability* of the atmosphere. Atmospheric stability plays a very important role in the dispersion of pollutants. It is closely tied to the way in which the temperature of the atmosphere varies with height. Plumes or parcels of air rising or sinking in the atmosphere tend to change temperature according to known physical principles. For example, a parcel of dry air will cool at a rate of $9.8\,°C\,km^{-1}$ whilst saturated air (in which condensation is occurring) cools at a slower rate close to $6.5\,°C\,km^{-1}$.

If air that is rising and cooling remains warmer than surrounding (ambient) air at the same level it is less dense and thus positively buoyant. This is called an *unstable* atmosphere and resulting conditions will promote vertical mixing. The opposite case, where the air parcel remains cooler than the surrounding atmosphere results in a *stable* atmospheric condition. It is associated with suppressed vertical mixing. For example, daytime heating of the surface produces an unstable mixed layer in the atmosphere that is typically 1 km deep. The top of the mixed layer is capped by a relatively shallow layer in which temperature increases with height (such a layer is very stable and is called an inversion as it represents the opposite of the typical cooling with height observed in the atmosphere). The presence of this capping inversion generally prevents pollutants from being mixed higher into the atmosphere. At night, the lower atmosphere is often stable (due to surface radiational cooling) and weak

Lake Superior College Library

mixing is confined to a shallow layer (often producing high pollutant concentrations). Consequently, pollutants may be dispersed into a much smaller volume than during the day time. In extreme cases temperatures may increase with height close to the surface creating very stable conditions. This type of surface temperature inversion is often associated with very poor air quality. To summarise, in stable layers ambient temperature decreases slowly with height (or may increase in the case of inversions) so turbulence and the diffusion of pollutants are suppressed. In unstable layers where ambient temperature decreases rapidly with height, turbulence and the diffusion of pollutants is enhanced.

Meteorological parameters also play an important role in governing local pollutant dispersion pathways, the formation of pollutant 'hotspots' and determining the residence time of pollutants in urban atmosphere. Thus understanding local, urban and regional scale variations in the thermal and dynamic characteristics of the urban atmosphere is fundamental to a comprehensive evaluation of air pollution in urban areas.

2   The Polluted Urban Atmosphere

Urban air pollution is a consequence of processes and phenomena operating at a range of spatial and temporal scales. A schematic sampling of these processes and phenomena is shown in Figure 2 and provides the structure for the remainder of this review. In summary, at scales ranging from individual buildings and street canyons to the entire city, microscale mechanical and thermally driven turbulence dominates local dispersion processes. However, these processes operate within a hierarchy of larger scales which provide the background state of the atmosphere that modulates air quality within urban areas. Hence, the passage of a synoptic scale cyclonic storm with precipitation, cloudiness, cool temperatures and strong winds will likely result in improved air quality in the urban context due to the dominance of deposition processes and good ventilation. In contrast, anticyclonic conditions may establish the conditions favourable for limited dispersion. In considering Figure 2, it is also important to note that a similar figure could be developed for chemical processes. That is, chemical transformations also have characteristic spatial and temporal scales For example, fresh emissions from traffic undergo rapid transformations (of order seconds to minutes) in the vicinity of highways, while ozone (O_3) formation in photochemical smog occurs on timescale of the order of hours and, as a consequence of horizontal transport by winds (advection), often has peak daily values some distance downwind of the urban source of precursor pollutants.[5] Finally, pollutants advected into urban areas as a consequence of long or medium range transport are composed of both secondary pollutants formed within the atmosphere and long-lived or "aged" chemical species.

In focusing on air quality in the urban atmosphere, it is important to recognise the meteorological consequences of the built form and human

Lake Superior College Library

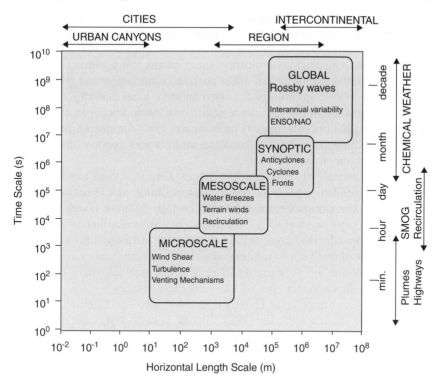

Figure 2 Temporal and spatial scales affecting atmospheric dispersion in the urban environment.

activities associated with the urban environment.[6] Thus, urban areas may experience quite different meteorological conditions than surrounding rural areas. This can have an impact on pollutant dispersion and atmospheric chemistry. Although urban surfaces vary considerably in form and composition around the globe, they are recognised to produce distinctive climates at a variety of temporal and spatial scales[7] and it is possible to draw generalisations about the urban impact on the atmosphere.[8,9]

One of the most pronounced effects of urbanisation on climate is the so called 'urban heat island' effect (UHI). This refers to the increased temperatures of both the urban surface and atmosphere compared to their rural counterparts.[10] The UHI is a product of increased heat storage capacity, enhanced radiative trapping and thermal inertia of anthropogenic materials, as well as local anthropogenic heat production.[10] The UHI phenomenon can have a direct impact on pollutant concentrations compared to surrounding rural areas by increasing instability within the atmosphere, and creating deeper mixing layers which enhance dispersion. The warm rough urban surface can also result in changes to mesoscale wind flow patterns which affect atmospheric dispersion.

Urban modifications to the atmosphere can also affect pollutant concentrations indirectly due to the influence of meteorological conditions on chemical processes. Temperature can affect the characteristics and rate of secondary chemical reactions which take place in the atmosphere. For example, an increase in air temperature of just 1 °C may result in a 14% increase in surface ozone concentrations during the summer in London, England.[11] Changes in relative humidity can affect particle composition and growth rates. Finally, atmospheric pollutants can also modify local climates acting as a feedback loop determining the nature of the relationship between meteorology and air pollution concentrations. For example, high concentrations of atmospheric pollutants (especially particulate matter) can modify the radiation balance and the hydrological cycle.[12,13]

Due to the complexity of the urban atmosphere it is useful to look at the relations between urban meteorology and air pollution at a variety of different scales. The urban atmosphere can be divided into different layers (Figure 3). The urban canopy layer (UCL) is that part of the atmosphere which occurs below roof tops. In this layer the atmosphere is strongly influenced by the characteristics of individual buildings and local building materials. Pollutant concentrations are highly variable in time and space and may be strongly influenced by the characteristics of local emission sources.

The urban boundary layer (UBL) is divided into the roughness sublayer (RSL) and the inertial sublayer (ISL) (Figure 3). The RSL extends from the ground to about 2.5 times the mean height of the buildings. Turbulent flows of heat and momentum have been shown to be intermittent and non-stationary in

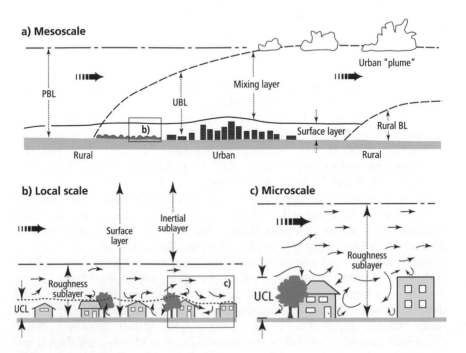

Figure 3   The structure of the urban atmosphere at a) mesoscale, b) local scale and c) microscale.[7]

this region, but little is known about the resulting pollutant concentrations.[9,14] In the ISL the atmosphere is more homogeneous. Turbulent flows are well developed and there are few local sources of pollutants, thus it is expected that pollutants are well mixed in time and space. A coherent urban plume may also be identified hundreds of kilometres (km) downwind of urban areas and thus impact the processes determining air pollution concentrations at regional and potentially global scales.[15]

The meteorological controls on pollutant concentrations and processes will be discussed at each of these scales in the following sections.

3 The Urban Canopy Layer

In urban areas, people spend a substantial component of their outdoor time in urban areas near busy roadways and intersections while commuting to work (on bicycles or foot), employed at local shops or cafes or using the pavement space for retail or recreational activities.[16] Although individuals may not remain in this environment for more than a few hours each day, the prevalence of local pollutant hot spots often results in significant exposure to pollutants. Further, in order to prepare for accidental or terrorist releases of biological, chemical or radioactive materials, it is important to understand the relationship between emissions characteristics and dispersal processes in determining concentrations near the surface at these scales.[17]

3.1 Emissions, Intra-Urban Variability and Data Sources

The temporal and spatial heterogeneity of emission patterns and microclimate combined with the complexity of the urban surface results in complex dispersion pathways at local scales within urban areas. This can lead to strong gradients in vertical and horizontal pollutant concentration.[18] Pollutant measurement sites are often classified using terms such as 'kerbside', 'roadside' or 'urban background' reflecting decreasing proximity from the primary source. Concentrations of primary pollutants decrease markedly within 150–300 m of a road source.[19,20] We would therefore expect the relation between traffic parameters and primary pollutants to be the strongest at kerbside sites. However, even at kerbside locations this relationship breaks down under congested conditions.[21] This suggests chemical and dispersion processes play an important role in determining local concentrations and may influence both the magnitude and timing of ambient pollutant peaks and cycles. Indeed, it has been noted that except at the heaviest traffic sites (such as Marylebone Road in Central London) the influence of meteorological controls on pollutant time series may be sufficient to modify or even mask all but the largest changes in local emissions characteristics.[3] This makes it very difficult to evaluate the success of pollutant abatement strategies based on information from pollutant time series.

Generally, urban background measurements are used to establish correlations between meteorological variables (often measured several kilometres away) and pollutant concentrations at hourly, daily and annual scales. However, there remains considerable debate about the intra-urban homogeneity of pollutant concentrations due to variations in source strength, meteorology, topography and location of monitoring sites.[22] For particulate matter, which represents a diverse range of particles arising from different sources and processes, the heterogeneity of the urban pollutant distribution is largely a function of particle size.[22] Coarse particles ($>2.5\,\mu m$ in diameter), with their large settling velocities and localised sources tend to show heterogeneous distribution, whilst smaller particles (0.1–$2.5\,\mu m$ diameter) with distributed sources and longer atmospheric residence times tend to show a more homogeneous distribution. Ultrafine particles (*e.g.* those emitted in motor vehicle exhaust) with their short lifetime tend to have highest concentrations near their primary source (often roadways). Measurements of pollutant concentrations in urban areas (which are often made within 2 metres of the surface) may be strongly influenced by local emission and flow characteristics and may not be representative of conditions elsewhere in the city. Thus fixed monitoring stations may not provide a good indication of personal exposure to pollutants, thereby hindering the establishment of strong relations between pollutant concentrations and meteorology[23] or health outcomes.[24,25]

3.2 Flow Patterns within the Urban Canopy Layer

At local scales the height to width ratio of buildings plays an important role in determining the local modification to wind flow patterns.[26] Where the buildings are a long way apart, the buildings act in isolation to modify flows (isolated roughness flow). As the buildings get closer together the wakes from one structure interfere with the patterns produced by surrounding structures (wake interference flows). Finally the obstacles become so close together that the majority of the flow skims over the surface and there is limited penetration from wind flows above into the urban canopy layer. Thus in this latter regime pollutant dispersal near the surface becomes limited as the UCL can be effectively decoupled from the UBL. Due to the complexity of the urban surface and the wide variety of model types available (each with their own limitations) the derivation of a general set of universal rules or processes relating pollutant dispersion to geometry or wind flow patterns remains elusive except at the most general scales.[4]

3.3 Street Canyons

Urban geometries are rarely uniform and consistent, and the scale of study becomes important. At local scales, flows are influenced by small variations in building morphology and it is very difficult to predict pollutant concentrations accurately. However at larger scales it is possible to move towards more generalised and predictable flow patterns. For example, the concept of a 'street canyon' or road flanked by buildings on both sides can be defined and is recognised as the standard unit of reference at local scales in urban studies from both measurement and modelling perspectives.[27]

When wind flows are close to perpendicular to the street canyon a spiral or helical type flow develops within the canyon. Wind flows cross the street canyon, hit the opposite (windward) wall and become entrained down the wall into the street canyon. Flows then cross the bottom of the street canyon and are forced back up the (leeward) wall closest to the prevailing flow forming a vortex (see Figure 3). This can result in the build up of pollutants on the leeward side of the street as flows sweep traffic pollutants across the road in one direction. Although the persistence of the vortex maybe intermittent,[28] field studies consistently report increased concentrations of primary pollutants on the leeward side of the road.[29] Increases of at least twice those on the windward side for background winds with a strong perpendicular component[30] and up to five times urban background concentrations[31] are reported. Thus the spatial variation of pollutants within the street canyon can be significant due to the presence or absence of vortex.

The orientation (with respect to wind direction and incoming solar radiation), height to width (H:W) ratio, length (distance between intersections), and symmetry of the canyon influence the characteristics of the flows. For example, when wind flow regimes are parallel to street canyons channelling flows funnel pollutants horizontally within the road network.[30,32] If the height to width ratio is large, multiple vortices may form, which further limits dispersion. Pollutant concentrations may build up within the canyon as pollutants remain entrained in the vortices and cannot escape out of the canyon. Although an exponential decrease in concentrations is generally reported with height,[33] small scale variations in geometry, distance from the canyon walls and wind flow regimes, and chemical reactivity of the pollutant can affect observed trends.[18,34]

Numerous studies have tried to measure flows and dispersion in urban canopies. Resource intensive field campaigns are necessary to produce repeatable results and draw conclusions which are transferable to other urban environments. Field measurements must be carefully designed and undertaken in areas of uniform urban geometry within areas of homogeneous fetch and ideally remain in place over periods of a year or more. The formation and characteristics of flow patterns in urban areas and their impacts on pollutant concentrations are perhaps easiest to determine using numerical modelling techniques and wind tunnel modelling techniques.[34–38] These have yielded some important insights.

Many questions remain about the exact conditions determining vortex development within a street canyon, its characteristics and strength[28] and the subsequent implications for pollutant dispersion. This is largely due to the details of the canyons (balconies, roof shapes, presence of cars or trees, intersections *etc.*) which influence thermal and mechanical dispersion processes at local scales in the real world.[39] For example, until recently, modelling studies typically neglected thermal influences on flows in the urban canopy. Thermal differences within the street canyon develop as a result of uneven patterns of radiation receipt and differences in building materials. They have the potential to affect the characteristics of vortex development[40–42] and the formation of secondary pollutants in the urban environment.[43,44]

The impact of vegetation, especially trees, on pollutant dispersion at a local scale is poorly understood. In one of a limited number of studies wind tunnel measurements are used to demonstrate that, depending on the height, width and density of the crown, the presence of trees within a street canyon can affect dispersion.[45] Trees close to the side of buildings limit vortex development and the reduced vertical flows were associated with increased pollutant concentrations on the leeward side of the canyon.

Traffic-induced turbulence may also be important in determining pollutant dispersion pathways, especially under very calm conditions.[46,47] This includes mechanical turbulence produced by vehicle motion as well as the thermal characteristics of the plume. For example the inclusion of traffic-induced turbulence in operational air quality models reduced the rate of model over-prediction under low wind speeds during extreme pollution events.[4]

Street canyons do not exist in isolation. The heterogeneity of urban areas may aid dispersion.[48,49] Recent studies have shown that street intersections can play an important role in the dispersion of pollutants.[32,50] Significant pollutant exchange can occur between streets at intersections.[50] Even small variations in the symmetry of the canyon or local wind flow patterns can result in quite different dispersion patterns. Complex wind flow patterns around buildings and street furniture can cause pollutants to accumulate in corner vortices or local eddies in intricate patterns.[26] Given that emission patterns also become more complex at intersections, as traffic flows become congested and speeds variable, it is very difficult to predict local concentrations at this scale.

3.4   Vertical Exchange Processes: Coupling the UCL to the UBL

Although the concentrations of pollutants within a street canyon are strongly affected by emissions characteristics, horizontal advection and dispersion within the street canyon network, the efficiency of vertical exchange processes which couple the air in the street canyon to the boundary layer aloft is also important.[51] However, to date there have been only a few studies directed towards modelling or measuring these exchange processes in the field.[4]

Modelling and measurement studies suggest that pollutants escape from the top of the street canyon primarily as a result of turbulent processes.[52,53] Exchange is thought to be driven by the penetration of turbulent structures from the urban boundary layer into the street canyon.[4,54,55] The efficiency of this process is thought to be strongly related to the height to width ratio of the street canyon.[56] Modelling studies have also suggested that vertical exchange processes increase as the angle of the incident wind flow increases relative to the street geometry and may be enhanced at intersections.[57]

Vertical exchange processes however may be reduced by the presence of trees with large crowns in street canyons, especially if they are taller than building heights.[45]

Thermal processes have also been shown to be important in determining vertical exchange processes. During the nocturnal period the release of heat stored within the urban canyon fabric can be sufficient to maintain weakly unstable mixing conditions over urban areas at night.[53,58–60] Thus under calm conditions the observation of coherent plumes of warm dirty air in the urban boundary layer at night are consistent with vertical exchange of heat and pollutants from the UCL.[53] Although daytime conditions have not been explicitly studied, the dominance of sensible heat fluxes from roof surfaces rather than canyon walls suggests that thermal processes may be less important during the day.[53,59] From an air pollution perspective, it is important to note that although pollutants may be transported out of street canyons, pollutants from the boundary layer aloft may also be transported downwards towards the surface.

At larger scales the increased roughness of the urban surface and higher surface temperatures increases the instability of the atmosphere compared to surrounding rural areas. This promotes mixing and increases the depth or volume of the boundary layer. Urban boundary layers show a strong diurnal cycle in character with stronger well-developed turbulence observed during the day. However, unlike their rural counterparts, the UBL often remains weakly unstable at night even under clear-sky, anticyclonic conditions. This is due in part to the release of heat stored within the urban fabric and in part to the increased roughness of the surface driving entrainment of warm air from aloft. This promotes mixing at night and inhibits the development of a stable layer.

Weakly unstable nocturnal boundary layers can have mixed consequences for surface air quality. Whilst surface emissions of pollutants are potentially diluted through a greater volume, enhanced mixing may also result in roof top emissions from chimneys becoming mixed downward towards the surface. Modelling studies also demonstrate that some pollutants (such as O_3) may be introduced into the urban canopy layer as a result of these vertical exchange processes.[43,44] However this phenomenon has yet to be widely reported in field studies.

4 Larger Scale Processes Affecting Urban Air Pollution

Air pollution processes and concentrations in urban areas are not completely controlled by the state of the atmosphere as dictated by the urban environment itself. Day-to-day variations in urban air quality are modulated in large part by processes operating at scales greater than the urban environment. These meteorological processes extend from the regional scale in which thermally and mechanically-forced flows dominate (*e.g.* sea breezes, winds channelled by terrain), to the synoptic scale (cyclonic and anticyclone systems that give our day-to-day weather), and finally to the hemispheric and global scales which permit long-range transport of pollutants and whose patterns in turn modulate synoptic weather patterns at the inter-annual and decadal scales (*e.g.* El Nino southern Oscillation or the North Atlantic Oscillation).

4.1   Mesoscale Flows

The interaction between regional scale wind flows driven by geographical features such as land–sea breezes or mountain–valley winds and urban areas can have an important effect on the dispersion of pollutants both within the urban area and down-wind of it. This is particularly true of cities located in regions of complex coastal terrain where the combined effects of water (lake or sea) breezes, terrain-forced flows (*e.g.* slope and valley winds) and low thermal internal boundary layers (TIBL) can contribute to the recirculation of pollutants.[61] For example, concentrations of particulate matter, sulfur dioxide and nitrous oxides were observed to decrease before the arrival of the sea-breeze front in Osaka and Kyoto.[62] Where these winds are circulatory in character, pollutant concentrations often increase with time as pollutants emitted the day before re-enter urban areas the following day. The strength and characteristics of these mesoscale flows may be intensified by urban effects.[63]

Due to the roughness of urban environments, wind flows may be slowed and change direction in urban areas. Consequently, mesoscale wind flow patterns may be changed as a result of the presence of the urban area.[4] Theoretical and modelling studies suggest that the development of the urban heat island could result in the development of a mesoscale circulation within the urban area with convergence of wind into the city centre and vertical motions circulating upwards and outwards.[64] However, in practice this has proved very difficult to observe largely due to weak flows and heterogeneity of the landscape. Nevertheless the 'bending' or deflection of large scale wind flows around urban areas has been documented.[65]

There remains some controversy as to whether the increased atmospheric instability and mixing prevalent over urban areas, especially large metropolitan areas, is sufficient to enhance precipitation, thunderstorm activity or storm frequencies.[15] This is in part due to difficulties in observing and predicting urban climate modification at this scale, but also due to the sensitivity of the processes to local parameters. For example, during calm conditions and light regional flow regimes the UHI may be sufficient to initiate storm activity which may enhance pollutant dispersion over large conurbations.[66,67] However, under stronger regional scale flows thunderstorms initiated outside the city tend to bifurcate and move around the city, limiting dispersion.[67]

Finally, the impact of mesoscale flows such as the sea breezes is often manifested in the urban context by complex layering of pollutants in the vertical. Numerous studies in locations such as the Los Angeles Basin, Mexico City, Tokyo, Athens, Vancouver and Chicago have revealed the presence of elevated layers of pollution formed by a variety of processes that ultimately contribute to both the "venting" of pollutants from the urban boundary layer

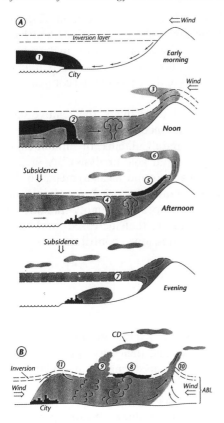

A Coastal and Orographic Effects
1. Offshore advection

2. Lofting at sea/lake breeze front
3. Advective venting

4. Undercutting of mixed layer by
   Sea/lake Breeze
5. Injection into layers by slope flows
6. Mountain venting

7. Evening stabilization

B Convection

8. Convective debris (CD)
9. Cloud venting
10. Low level convergence of winds
11. Urban Heat Island (UHI) effect

Figure 4 Schematic depiction of processes of elevated layer development and boundary layer venting over urbanised (a) complex coastal terrain and (b) flat inland areas.

and the "recirculation" of pollutants.[68–70] These processes and mechanisms are shown schematically in Figure 4.[71]

4.2 Regional Transport

In many parts of the world urban areas are downwind of other large conurbations. Thus regional scale transport of pollutants into urban areas can make a significant contribution to ambient levels. For example, high concentrations of PM_{10} and $PM_{2.5}$ in London are frequently associated with regional scale transport of secondary particulates and their precursors from continental Europe.[72,73]

However, conditions which promote large scale regional transport of ambient pollutants (anticyclonic conditions characterised by high atmospheric pressure at the surface, light wind speeds and limited mixing volumes) are also associated with periods when dispersion of local pollutants is poor.

For example, highest concentrations of particulate matter in London, England have been observed to occur during stable atmospheric conditions and light winds.[72] Thus one of the biggest challenges facing air quality managers is determining how much of the ambient air pollution measured in urban areas is transported into the urban area (and thus beyond local controls) compared to emissions from local pollutant sources.

In theory, atmospheric models can be used to evaluate the relative contribution of local pollutants.[74] However, the success of this technique is dependent on both the accuracy of the model and availability of suitable input data (detailed vehicle emissions and complex wind field analysis).[3] As a consequence, many studies use rural or urban background measurements to distinguish between local and regional scale pollutant sources.[33] For example, the number of times PM_{10} concentrations exceed European air quality standards in London is higher at road-side sites, demonstrating that increased local emissions and reduced dispersion of those emissions contribute to high concentrations.[73]

4.3   Synoptic Scales

In contrast to mesoscale circulations whose impact is felt in urban settings over the diurnal timescale, the emission of pollutants and their subsequent chemical transformation, transport and dispersion in urban settings is modulated on timescales of days by synoptic patterns that dictate our "weather". For example, in mid-latitudes, the passage of transient anticyclones and depressions (with their markedly different characteristics with respect to cloudiness, precipitation, wind and radiation) strongly influences the air quality of urban areas. Whereas the passage of cyclonic systems usually promotes good air quality because of the impact of strong winds, precipitation (rainout and deposition), low temperatures and high mixing depths, anticyclonic systems tend to reduce air quality because of light winds (low ventilation) and reduced mixing depths due to subsidence. Increased temperatures and solar radiation associated with high pressure systems also promote photochemistry and lead to smog situations that are well-documented in locations such as Los Angeles, or in north eastern USA where the "back of the high" synoptic weather pattern is responsible for severe regionally degraded air quality.[75] Recently, application of contemporary synoptic climatological techniques to the analysis of relations between synoptic scale pressure patterns and atmospheric pollution in ten cities in Northern China has shown that air quality is degraded during anticyclonic conditions and into the prefrontal period. Thereafter, post-frontal air quality significantly improves.[76]

Synoptic patterns also create the circumstances under which mesoscale circulations (described above) can form. For example, anticyclonic conditions, which promote poor air quality as a result of low ventilation, also promote the development of thermally forced circulations such as sea breezes. These may ultimately lead to further degradation of air quality due to recirculation processes described above. [61,71]

Finally, day-to-day and seasonal weather patterns are instrumental in influencing urban emission patterns associated with space heating (and cooling) as well as commuter transportation. For example, stormy winter days may prompt increases in domestic space heating and automobile usage whilst warm anticyclonic days may encourage bicycling and curtailment of space heating.

4.4 Global Scales and "Chemical Weather"

The events of Chernobyl 1986 highlighted the extent to which pollutants may be transported around the globe in the mid-latitude westerlies. Ultimately, that event affirmed that even at the urban scale, pollutants from "afar", whether regionally or globally, may contribute significantly to the "background" of pollutants in the urban setting. Most importantly, these pollutants advected into an urban area from beyond may be sufficient to significantly influence local air quality and indeed contribute significantly to exceedences of local air quality standards. For example, it has recently become apparent that intercontinental transport of pollutants has a significant impact on western North America.[77] On a monthly basis, Asian crustal dust contributes about $0.6-1.6 \, \mu g \, m^{-3}$ to the concentrations along the west coast (where mean values may be the order of $5-6 \, \mu g \, m^{-3}$. For individual transport events, the effects may be much more significant as in the case of the 1998 and 2001 Asian dust storms which noticeably degraded air quality across the North America. Furthermore, it has been demonstrated that a plume from a Siberian forest fire contributed 15 ppb to an 8 h average O_3 concentration of 96 ppb in the Seattle region. This additional O_3 burden was sufficient to place air quality above the National Ambient Air Quality Standards.[77]

The effects of regional and long-range transport of pollutants is now a well recognised and researched global issue and has spawned the term "*Chemical Weather*" that reflects the increasing importance and appreciation of the "local, regional and global distributions of important trace gases and aerosols and their variability on time scales of minutes to hours to days, particularly in light of their various impacts, such as on human health, ecosystems, and the meteorological weather and climate".[78]

Finally, at the global scale, the presence of inter-annual and decadal variability, as manifested in the El Nino Southern Oscillation, the Pacific Decadal Oscillation and the North Atlantic Oscillation, amongst others, provide low frequency background variability that modulates synoptic patterns and ultimately air quality in cities over longer timescales. It is therefore important to consider such cycles when examining trends in air quality in urban settings.

5 Conclusions

Although *inter-* and *intra*-urban variations in pollutant emissions are important, the state of the atmosphere clearly plays a key role in determining air quality in urban areas at a variety of temporal and spatial scales.

Meteorological processes directly affect the rate at which pollutants accumulate and disperse in the atmosphere and may have an additional indirect effect on pollutant emission rates and chemical transformations.

Generally, as the ability of the urban atmosphere to disperse pollutants horizontally and vertically increases, pollutant concentrations decrease. However, complex feedback loops exist between pollutant emission rates, urban form and meteorological processes, local and regional pollutant transport rates and atmospheric chemistry such that individual parameters are rarely independent of each other. Thus the identification of relations between meteorological parameters and pollutant concentrations applicable in urban environments around the world remain elusive at all but the most general scales.

As urban expansion continues we can expect a further deterioration in urban air quality. Global scale changes in climate can also be expected to affect atmospheric composition at a variety of scales. However, improved understanding of the links between meteorology and air quality can provide insights into the design of cities to ensure that future urban expansion optimises natural ventilation and minimizes population exposure.

Acknowledgements

Figure 1 prepared by Igor Drecki, Geo-graphics Unit manager, School of Geography, Geology and Environmental Science, University of Auckland.

References

1. World-Bank, 2007, Accessed online September 2008 from Table 3.13, pp. 174–175, http://siteresources.worldbank.org/DATASTATISTICS/Resources/table3_13.pdf
2. R. Colvile, E. J. Hutchinson, J. S. Mindell and R. F. Warren, *Atmos. Environ.*, 2001, **35**(9), 1537–1565.
3. D. C. Carslaw and N. Carslaw, *Atmos. Environ.*, 2007, **41**(22), 4723–4733.
4. R. E. Britter and S. R. Hanna, *Annu. Rev. Fluid Mech.*, 2003, **35**, 469–496.
5. J. A. Salmond and I. G. McKendry, *Atmos. Environ.*, 2002, **36**(38), 5771–5782.
6. A. J. Arnfield, *Int. J. Climatol.*, 2003, **23**, 1–26.
7. T. R. Oke, in *Surface Climates of Canada,* ed. W. G. Bailey, T. R. Oke and W. R. Rouse, McGill-Queen's University Press, Montréal, 1997, 303–327.
8. C. S. B. Grimmond and T. R. Oke, *J. Appl. Meteorol.*, 1999, **38**(9), 1262–1292.
9. M. Roth, *Q. J. R. Meteorol. Soc.*, 2000, **126**(564), 941–990.
10. T. R. Oke, *Q. J. R. Meteorol. Soc.*, 1982, **108**(455), 1–24.
11. D. Lee, *Geography*, 1993, **78**, 77–79.

12. R. K. Kaufmann, K. C. Seto, A. Schneider, Z. Liu, L. Zhou and W. Wang, *J. Climate*, 2007, **20**(10), 2299–2306.
13. A. Givati and D. Rosenfeld, *J. Appl. Meteorol.*, 2004, **43**(7), 1038–1056.
14. M. Roth, J. A. Salmond and A. N. Satyanarayana, *Boundary Layer Meteorol.*, 2006, **121**(2), 351–375.
15. P. J. Crutzen, *Atmos. Environ.*, 2004, **38**(21), 3539–3540.
16. S. Kaur, M. J. Nieuwenhuijsen and R. Colvile, *Atmos. Environ.*, 2007, **41**(23), 4781–4810.
17. K. J. Allwine, J. Shinn, G. E. Streit, K. L. Clawson and M. Brown, *Bull. Am. Meteorol. Soc*, 2002, **22**, 521–536.
18. M. Vakeva, K. Hameri, M. Kulmala, R. Lahdes, J. Ruuskanen and T. Laitinen, *Atmos. Environ.*, 1999, **33**(9), 1385–1397.
19. Y. F. Zhu, W. C. Hinds, S. Kim and C. Sioutas, *J. Air Waste Manage. Assoc.*, 2002, **52**(9), 1032–1042.
20. N. L. Gilbert, M. S. Goldberg, B. Beckerman, J. R. Brook and M. Jerrett, *J. Air Waste Manage. Assoc.*, 2005, **55**(8), 1059–1063.
21. E. L. Agus, D. T. Young, J. J. N. Lingard, R. J. Smalley, J. E. Tate, P. S. Goodman and A. S. Tomlin, *Sci. Total Environ.*, 2007, **386**, 65–82.
22. J. D. Wilson, S. Kingham, J. Pearce and A. Sturman, *Atmos. Environ.*, 2005, **39**(34), 6444–6462.
23. R. M. Harrison, A. M. Jones and R. Barrowcliffe, *Atmos. Environ.*, 2004, **38**(37), 6361–6369.
24. H. S. Adams, M. J. Nieuwenhuijsen, R. N. Colvile, M. A. S. McMullen and P. Khandelwal, *Sci. Total Environ.*, 2001, **279**(1–3), 29.
25. J. Gulliver and D. J. Briggs, *Atmos. Environ.*, 2004, **38**(1), 1.
26. T. R. Oke, *Energy Build.*, 1988, **11**, 103–113.
27. T. R. Oke, *Theor. Appl. Climatol.*, 2006, **84**(1–3), 179–190.
28. I. Eliasson, B. Offerle, C. S. B. Grimmond and S. Lindqvist, *Atmos. Environ.*, 2006, **40**(1), 1–16.
29. S. Rafailidis, *Int. J. Environ. Pollut.*, 2000, **14**(1–6), 538–546.
30. J. W. Boddy, R. J. Smalley, N. S. Dixon, J. E. Tate and A. S. Tomlin, *Atmos. Environ.*, 2005, **39**(17), 3147–3161.
31. F. Palmgren and K. Kemp, *NERI*, Roskilde, Denmark, 1999.
32. A. Dobre, S. J. Arnold, R. J. Smalley, J. W. Boddy, J. F. Barlow, A. S. Tomlin and S. E. Belcher, *Atmos. Environ.*, 2005, **39**(26), 4647–4657.
33. S. Vardoulakis, B. E. A. Fisher, K. Pericleous and N. Gonzalez-Flesca, *Atmos. Environ.*, 2003, **37**(2), 155–182.
34. P. Kastner-Klein and E. J. Plate, *Atmos. Environ.*, 1999, **33**(24–25), 3973–3979.
35. A. J. Arnfield, *Prog. Phys. Geogr.*, 2005, **29**, 426–437.
36. P. Kastner-Klein, R. Berkowicz and R. Britter, *Meteorol. Atmos. Phys.*, 2004, **87**(1–3), 121–131.
37. A. J. Arnfield, *Prog. Phys. Geogr.*, 2001, **25**(4), 560–569.
38. M. Pavageau and M. Schatzmann, *Atmos. Environ.*, 1999, **33**(24–25), 3961–3971.
39. S. Rafailidis, *Int. J. Environ. Pollut.*, 2001, **16**(1–6), 393–403.

40. J. J. Kim and J. J. Baik, *Atmos. Environ.*, 2001, **35**(20), 3395–3404.
41. J. J. Kim and J. J. Baik, *Adv. Atmos. Sci.*, 2005, **22**(2), 230–237.
42. B. Offerle, I. Eliasson, C. S. B. Grimmond and B. Holmer, *Boundary Layer Meteorol.*, 2007, **122**(2), 273–292.
43. J. Baker, H. L. Walker and X. Cai, *Atmos. Environ.*, 2004, **38**(39), 6883–6892.
44. J. J. Baik, Y. S. Kang and J. J. Kim, *Atmos. Environ.*, 2007, **41**(5), 934–949.
45. C. Gromke and B. Ruck, *Atmos. Environ.*, 2007, **41**(16), 3287–3302.
46. P. Kastner-Klein, R. Berkowicz and E. J. Plate, *Int. J. Environ. Pollut.*, 2000, **14**(1–6), 496–507.
47. P. Kastner-Klein, E. Fedorovich, M. Ketzel, R. Berkowicz and R. Britter, *Environ. Fluid Mech.*, 2003, **3**(2), 145–172.
48. A. T. Chan, W. T. W. Au and E. S. P. So, *Atmos. Environ.*, 2003, **37**(20), 2761–2772.
49. A. T. Chan, E. S. P. So and S. C. Samad, *Atmos. Environ.*, 2001, **35**(24), 4089–4098.
50. A. Robins, E. Savory, A. Scaperdas and D. Grigoriadis, *Water, Air Soil Pollut. Focus 2*, 2002, 381–393.
51. F. Caton, R. E. Britter and S. Dalziel, *Atmos. Environ.*, 2003, **37**(5), 693–702.
52. J. J. Baik and J. J. Kim, *Atmos. Environ.*, 2002, **36**(3), 527–536.
53. J. A. Salmond, T. R. Oke, C. S. B. Grimmond, S. Roberts and B. Offerle, *J. Appl. Meteorol.*, 2005, **44**(8), 1180–1194.
54. A. Christen, E. van Gorselb and R. Vogt, *Int. J. Climatol.*, 2007, **27**, 1955–1968.
55. P. Louka, S. E. Belcher and R. G. Harrison, *Atmos. Environ.*, 2000, **34**(16), 2613–2621.
56. J. F. Sini, S. Anquetin and P. G. Mestayer, *Atmos. Environ.*, 1996, **30**(15), 2659–2677.
57. J. J. Kim and J. J. Baik, *Atmos. Environ.*, 2004, **38**(19), 3039–3048.
58. A. Christen and R. Vogt, *Int. J. Climatol.*, 2004, **24**(11), 1395–1422.
59. C. S. B. Grimmond, J. A. Salmond, T. R. Oke, B. Offerle and A. Lemonsu, *J. Geophys. Res. Atmos.*, 2004, **109**, 19.
60. B. Offerle, C. S. B. Grimmond, K. Fortuniak and W. Pawlak, *J. Appl. Meteorol. Climatol.*, 2006, **45**, 125–136.
61. D. G. Steyn, in Air Pollution Modeling and its Application XI, ed. S.E. Gryning, F.A. Schiermeier, Plenum Press, New York, 1996.
62. Y. Ohashi and H. Kida, *J. Appl. Meteorol.*, 2004, **43**(1), 119–133.
63. Y. Ohashi and H. Kida, *J. Appl. Meteorol.*, 2002, **41**(1), 30–45.
64. A. Lemonsu and V. Masson, *Bound. Layer Meteorol.*, 2002, **104**(3), 463–490.
65. R. Bornstein, *Modeling the Urban Boundary Layer*, American Meteorological Society, Boston, MA, 1987.
66. E. Jauregui and E. Romales, *Atmos. Environ.*, 1996, **30**(20), 3383–3389.
67. R. Bornstein and Q. Lin, *Atmos. Environ.*, 2000, **34**(3), 507–516.
68. A. Clappier, A. Martilli, P. Grossi, P. Thunis, F. Pasi, B. C. Krueger, B. Calpini, G. Graziani and H. van den Bergh, *J. Appl. Meteorol.*, 2000, **39**(4), 546–562.

69. P. Grossi, P. Thunis, A. Martilli and A. Clappier, *J. Appl. Meteorol.*, 2000, **39**(4), 563–575.

70. M. Yimin and T. J. Lyons, *Atmos. Environ.*, 2003, **37**(4), 443–454.

71. I. G. McKendry and J. Lundgren, *Prog. Phys. Geogr.*, 2000, **24**, 359–384.

72. A. Charron and R. M. Harrison, *Environ. Sci. Technol.*, 2005, **39**, 7768–7776.

73. A. Charron, R. M. Harrison and P. Quincey, *Atmos. Environ.*, 2007, **41**(9), 1960–1975.

74. D. J. Carruthers, H. A. Edmunds, A. E. Lester, C. A. McHugh and R. J. Singles, *Int. J. Environ. Pollut.*, 1998, **14**, 1–6.

75. K. C. Heidorn and D. Yap, *Atmos. Environ. Special Issue*, 1986, **20**(4), 695–703.

76. Z. H. Chen, J. B. Li, X. R. Guo, W. H. Wang and D. S. Chen, *Atmos. Environ.*, 2008, **42**(24), 6078–6087.

77. T. Keating, J. West and D. Jaffe, *Environ. Monitor.*, 2005, 28–30.

78. M. G. Lawrence, O. Hov, M. Beekmann, J. Brandt, H. Elbern, H. Eskes, H. Feichter and M. Takigawa, *Environ. Chem.*, 2005, **2**(1), 6–8.

Atmospheric Chemical Processes of Importance in Cities

WILLIAM BLOSS

ABSTRACT

Chemical processes affect the current and future composition of the atmosphere, controlling the abundance of a range of pollutants harmful to health, vegetation and materials. Atmospheric composition above a city is a function of the incoming air mass, emissions associated with the urban environment and chemical processing which affects the levels of primary pollutants and creates secondary species. In the city environment, local emissions and their short-timescale chemical processing tend to dominate atmospheric composition. Key atmospheric components include hydrocarbons and oxides of nitrogen emitted from vehicles and a range of domestic and industrial activities, and particulate matter produced mechanically and chemically within the atmosphere. In this article the key emissions and the dynamical environment within which the chemistry occurs are briefly introduced, followed by a description of the predominant chemical processes occurring in cities. In the gas phase the focus is upon the evolution of nitrogen oxides over the city scale, the role of atmospheric free radicals, nitrogen oxides and hydrocarbons in the production of ozone, and the different emission control regimes which are encountered. In the condensed phase, the origins, composition and chemistry of atmospheric particles found in urban environments are described, together with their interactions with the gas-phase component of the aerosol mixture. The chapter concludes by considering the limitations introduced by the chemical complexity of the urban environment, and some current uncertainties in our understanding of the chemistry occurring above our cities.

Issues in Environmental Science and Technology, 28
Air Quality in Urban Environments
Edited by R.E. Hester and R.M. Harrison
© Royal Society of Chemistry 2009
Published by the Royal Society of Chemistry, www.rsc.org

1 Introduction

The increasing urbanisation of our planet is causing the distribution of population to pass a major milestone during the present decade: more people, globally, now live in cities than in rural areas.[1] Local atmospheric pollution is one of the major environmental challenges facing urban populations, affecting the health and environment of potentially billions of people, particularly in the megacities (those with populations in excess of 10 million), which are growing most rapidly, and are predominantly located in nations amongst those with the least established regulations and technology to mitigate urban pollution. An understanding of the transformations of atmospheric constituents in the city environment is of importance to inform development of strategies to minimise pollution, in addition to its intrinsic interest.

The chemical processes found in the atmosphere within and above cities differ from those of the more remote boundary layer as a consequence of the magnitude and diverse and complex nature of local anthropogenic emission sources, the higher concentrations of many trace atmospheric species found as a result of this, and of the dynamic micro- and macro-environments in the urban landscape. In this article we introduce the principal emission sources found in the urban environment, briefly consider the dynamical processes which can affect the evolution and hence consequences of these emissions, and then describe the key atmospheric chemical processes which occur, highlighting some areas in which recent developments have identified gaps in our understanding, and considering interrelated roles of the gas and particulate phases.

2 Emissions in the City Environment

Air pollution issues in cities associated with the combustion of coal for domestic uses, and for industrial operations such as lime production, are noted in historical documents dating to at least the 13th century, when a Commission was established to address the problems associated with coal smoke in London. By the 16th–17th centuries, some of the earliest proposals for "emission control legislation" such as John Evelyn's *Fumifugium* of 1661 were drafted for (unsuccessful) presentation to Parliament.[2] Increasing domestic and industrial coal combustion led to the London smogs of the 19th and 20th centuries, which reached their nadir during the winter of 1952–53 when an estimated 4000 deaths are thought to have resulted from inhalation of the particulate and sulfuric acid-laden air. The principal emissions from coal combustion are particulate matter, soot (or black smoke) from incomplete combustion, and sulfur dioxide, with some production of nitrogen oxides through the Zeldovitch mechanism (see Equations 1 and 2). The introduction of clean-burning coal with a low sulfur content and the widespread use of electricity and natural gas for domestic heating and cooking have reduced the impact of coal combustion upon the atmospheric environments of cities in many countries, although in nations such as India and China, with rapidly expanding coal-fired power industries, substantial emissions remain. While heavy industrial enterprises have largely relocated away from city centres in many western nations, and have undergone extensive clean-up with, for example, the introduction of SO_2 abatement technologies for power station exhaust stacks, commercial/light industrial use of solvents, such as paints, adhesives, dry-cleaning solvents and aerosol propellants, are a substantial source of volatile organic compounds (VOCs) to the urban environment.

Exhaust emissions from internal combustion-engined vehicles play a key role in urban atmospheric processing. Gas-phase emissions from motor vehicles include NO_x (NO and NO_2), CO and VOCs. The introduction of three-way catalytic converters for petrol (spark ignition) engined vehicles, and oxidation catalytic converters together with exhaust gas recirculation for diesel (compression ignition) vehicles, has greatly reduced emissions of CO, VOCs and NO_x in many cities; however, regions outside the developed world suffer from much higher emissions due to the older technologies still prevalent (and legislation-driven displacement of vehicle fleets from elsewhere). An indication of the range of emissions encompassed by different vehicle type and emissions control technologies may be gleaned from the emission factors summarised in Table 1. NO_x at the tailpipe is conventionally assumed be partitioned (on average) into 95% NO and 5% NO_2, although there is some evidence that the direct fraction of NO_2 may be greater than this, and increasing across the fleet on average (see later sections).

In the case of the UK, the transport sector dominates emissions of CO (53%) and NO_x (45%), but is (according to current inventories) less dominant for non-methane hydrocarbons (NMHCs, 14%),[3] although a further significant source of NMHCs to the urban environment is fugitive emissions from filling station activities. Some measurements (see below) suggest that emissions inventories may undercount the VOC releases associated with vehicle traffic, particularly for larger species such as aromatic compounds. Increasing use of alternative fuel formulations may alter the emissions speciation and levels; in particular biofuels such as ethanol are associated with increased emissions of carbonyl compounds such as formaldehyde and acetaldehyde, which are both toxic in their own right, and act as sources of oxidant radicals in the environment.[4]

Table 1   Selected Vehicle Emission Factors for the UK Fleet.

	CO	VOCs	NO_x
Petrol Car – Euro III	0.6	0.04	0.2
Petrol Car – Pre-legislation	25	2.4	2.2
Diesel Car – Euro III	0.2	0.04	0.5
Diesel Car – Pre-legislation	0.7	0.16	0.6
HGV – Euro III	2.3	1.0	9.7
HGV – Pre-legislation	4.0	4.2	20.7

Values are in $g\,km^{-1}$, and apply to a mid-sized petrol car (1.4–2 l), <2 l diesel, and articulated HGV, for a constant speed of 40 km h^{-1}. Data taken from the UK National Atmospheric Emissions Inventory.

Urban atmospheric environments in developed nations are dominated by traffic-related sources (primarily CO, NO_x and particulate matter) and light-industrial/commercial emissions. In developing nations, domestic processes (heating and cooking) may also be substantial contributors, particularly where wood is the primary fuel and levels of CO, soot and other particulate matter (emissions associated with incomplete/low temperature combustion) are more significant. The atmospheric environment above the city is therefore characterised by the properties of the inflowing air, upon which are superimposed emissions of NO_x, VOCs, and particulate matter, leading to atmospheric processing which affects the chemical composition on timescales of minutes to days.

3 Dynamic Considerations

While a detailed account of the dynamic and meteorological considerations which can affect atmospheric chemical processing and pollutant levels (air mass origin, wind speed, dispersion, precipitation, solar insolation, temperature) is beyond the scope of this paper, a brief consideration of the key dynamical features is useful to provide a context for the chemical discussion which follows (these factors are considered in greater detail in accompanying articles within this issue).

When meteorological conditions lead to a stable, isolated atmosphere above an urban region, the lack of dilution of emissions leads to enhanced atmospheric processing and pollution. An example of such behaviour is the occurrence of stagnant periods associated with wintertime high-pressure systems, with temperature inversions leading to isolation of the boundary layer airmass – conditions somewhat characteristic of many UK winters, including those of the London smog events of the 1950s.[5] The cold weather is conducive to increased heating-related emissions of NO_x and potentially SO_2, while these together with vehicle and industrial emissions are trapped, leading to very high NO_x levels. Such wintertime smog events, sometimes termed "classic" or "London" smogs, may be contrasted with the photochemical smog events now seen more commonly in summertime within and downwind of developed areas, and which result from the chemical processing of the VOC–NO_x mixtures in the presence of sunlight leading to the formation of secondary pollutants such as ozone, peroxyacetyl nitrate (PAN) and various hydrocarbon degradation products. Photochemical smog episodes are associated with good weather, *i.e.* light winds within high-pressure systems, and may be augmented by local conditions, such as the advection-driven temperature inversions found in Southern California, or landscape factors, such as the high ground nearly encompassing Mexico City.

On a more local scale, the mixing from street-level emissions into the wider boundary layer is influenced by the surface roughness or topographic variation, which can enhance mixing from above the street level into the overlying boundary layer. On the scale of individual streets, the street canyon formed from rows of buildings can temporarily isolate emissions from the overlying boundary layer, leading to the build up of pollutants. Wind flows within canyons can also affect pollutant dispersion, ranging from channelling of pollutants along the canyon axis, to recirculation within the canyon: if the overlying wind has a significant cross-canyon component, a vortex can be established within the canyon such that at ground level pollutant transport is towards the nominally upwind side.

Typical residence times within an urban environment are of the order of a few minutes (canyon) to a few hours (city-wide) to a day or so (wider conurbation); it is within these relatively short timeframes that atmospheric chemical processes can affect the composition of urban air.

4 Gas-phase Chemical Processing

4.1 NO_x–Ozone Interactions

Gas-phase processes in urban environments are dominated by NO_x, produced primarily from exhaust emissions associated with internal combustion engined vehicles. NO_x is formed in any high-temperature combustion process (performed in air) through the thermal decomposition of molecular oxygen, leading to a chain reaction producing NO (the Zeldovitch mechanism):

$$O + N_2 \rightarrow NO + N \tag{1}$$

$$N + O_2 \rightarrow NO + O \tag{2}$$

Conventionally, NO_x emitted from vehicles at the exhaust point is assumed to be in an approximately 0.95 : 0.05 (NO : NO_2) ratio, *i.e.* NO dominates (see comments below however). Directly following emission from the vehicle exhaust, NO levels may be sufficiently high that the normally unimportant NO recombination Reaction (3) becomes significant:

$$NO + NO + O_2 \rightarrow 2NO_2 \tag{3}$$

Rapid dilution reduces NO mixing ratios to below the level at which this process is important, and NO_x chemistry is dominated (in daylight) by the following three *photochemical steady state* reactions:

$$NO + O_3 \rightarrow NO_2 + O_2 \tag{4}$$

$$NO_2 + h\nu \rightarrow NO + O(^3P) \tag{5}$$

$$O(^3P) + O_2 + M \rightarrow O_3 + M \tag{6}$$

The (^3P) label denotes that the oxygen atoms formed in Reaction (5) do not have any excess energy and are in their *ground state* – O atoms in this state almost always react with molecular oxygen, O_2, to form ozone, O_3, in the atmosphere (Reaction 6). This is in contrast to higher-energy oxygen atoms, denoted $O(^1D)$, formed *via* different routes and which can react with atmospheric water vapour (see below). M in Equation (6) (and later equations) represents a third molecule such as nitrogen or oxygen which is involved in the collision, but is not changed by the chemical reaction.

The timescale of Reactions (4) – (6) is a few minutes or less under typical mid-latitude boundary layer daylight conditions, thus (in the absence of other factors) they define an equilibrium between NO, O_3 and NO_2, given by the Leighton Relationship.[6]

$$[NO_2]/[NO] = k_4[O_3]/j_5 \qquad (7)$$

In this expression, k_4 is a constant which describes the rate of the $NO + O_3$ Reaction (4) at a given temperature, while j_5 is the photolysis rate constant for Reaction (5), the rate at which NO_2 is destroyed by sunlight to form $NO + O$. The value of j_5 therefore depends upon solar radiation levels and varies with location, time, weather *etc.* Considering Reactions (4) – (6) alone, the mixing ratio of oxidant O_x (sum of O_3 and NO_2 mixing ratios) is conserved, with the partitioning between the component forms O_3 and NO_2 determined by the overall levels of NO_x, O_3 and solar intensity. This behaviour is illustrated in the measurement sequence shown in Figure 1, which shows the variation in the

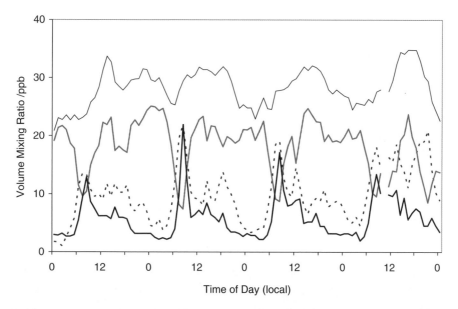

Figure 1 Mixing ratios of NO (heavy black line), NO_2 (dashed line), O_3 (grey line) and O_x (NO_2 and O_3) (fine black line) measured at a city-centre location in Birmingham, UK over a four-day midweek period during autumn 2008.

levels of NO, NO_2, O_3 and O_x measured at a city-centre location in Birmingham, UK, over a period of four week days. These data show the major peak in NO_x associated with the morning rush hour, with a lesser peak also sometimes apparent in the afternoons. The anticorrelation between NO_2 and O_3 as a consequence of Reaction (4) is clearly apparent, together with the near-constant level of total oxidant, O_x.

The meteorological conditions during the period of the observations shown in Figure 1 were such that the city-centre air was being steadily replenished by clean (low NO_x) airmasses from the wider region, with the result that no overall trend in ozone or O_x is discernable; however, atmospheric observations under different conditions, and experiments in photochemical simulation chambers, reveal that the processing of hydrocarbons in the presence of NO_x results in the production of ozone. Ozone production is a result of the interaction between peroxy radicals and NO_x.

4.2 Oxidant Radicals and Ozone Production

The gas-phase hydroxyl radical, OH, is the principal oxidant of the atmosphere, initiating the degradation of most organic compounds, ranging from globally distributed species such as methane and HCFCs, to local pollutants such as benzene. In the free troposphere, production of OH is driven by the solar fragmentation or photolysis of ozone at short wavelengths ($< \sim 310\,nm$), leading to the production of high energy (electronically excited) oxygen atoms, denoted $O(^1D)$, which have enough energy to be able to react with water vapour:

$$O_3 + h\nu \rightarrow O(^1D) + O_2 \tag{8}$$

$$O(^1D) + H_2O \rightarrow OH + OH \tag{9}$$

OH reactions with hydrocarbons (denoted below as RH) in the atmosphere lead to the production of organic peroxy radicals (in the case of hydrocarbon compounds) or hydroperoxy radicals (in the case of carbon monoxide):

$$OH + RH \rightarrow H_2O + R \tag{10}$$

$$R + O_2 + M \rightarrow RO_2 + M \tag{11}$$

$$OH + CO \rightarrow H + CO_2 \tag{12}$$

$$H + O_2 + M \rightarrow HO_2 + M \tag{13}$$

The fate of the peroxy radicals depends upon the atmospheric chemical environment: in clean environments, those low in NO_x, the peroxy radicals undergo self- and cross-reaction forming peroxides, alcohols and aldehydes, which generally have atmospheric lifetimes of several days and are removed by wet deposition; the overall effect of this reaction chain is therefore to destroy ozone through photolysis (Reaction 8) if NO_x levels are sufficiently low.

$$RO_2 + HO_2 \rightarrow ROOH + O_2 \qquad (14)$$

$$HO_2 + HO_2\ (+M) \rightarrow H_2O_2\ (+M) \qquad (15)$$

In the presence of NO_x, as is typical for an urban environment, radical propagation occurs: organic peroxy radicals react with NO, forming (primarily) NO_2 plus an alkyl radical. The alkyl radical reacts essentially instantaneously with oxygen, forming an aldehyde plus a hydroperoxy radical; this is turn also reacts with NO forming a further molecule of NO_2 and regenerating OH.

$$RO_2 + NO \rightarrow RO + NO_2 \qquad (16)$$

$$RO + O_2 \rightarrow R'CHO + HO_2 \qquad (17)$$

$$HO_2 + NO \rightarrow OH + NO_2 \qquad (18)$$

The NO-driven propagation reactions, summarised in Figure 2, therefore have the effect (at moderate levels of NO_x, for a given VOC loading) of

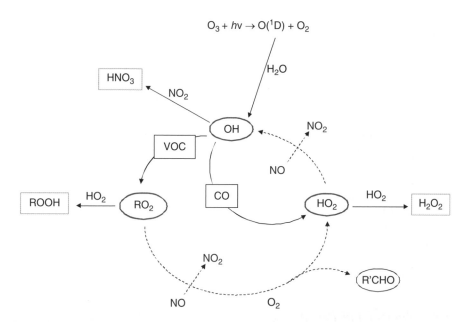

Figure 2 A HO_x-centric summary of hydrocarbon oxidation, showing the additional processes (dashed lines) which occur in the presence of NO_x. Reactions which are rapid and sequential in the boundary layer are combined.

increasing OH and hence the processing of VOCs, and of converting NO to NO_2. The subsequent solar photolysis of NO_2 leads to the production of O atoms and hence overall production of ozone. (The compensation point between ozone destruction and production is of the order of 5–25 ppt, very much lower than typical urban NO_x levels of a few ppb.)

The ozone production chemistry is limited by termination reactions which remove the reactive radical species, converting them to more stable compounds, such as the reaction of OH with NO_2 to form nitric acid (19), and the formation of organic nitrates as a minor channel (for small VOCs) of peroxy radical + NO reactions. Nitric acid (HNO_3) is the primary sink for NO_x in urban environments: it is highly soluble, and is readily lost to the condensed phase.

$$OH + NO_2 + M \rightarrow HNO_3 + M \tag{19}$$

The formation and fate of peroxyacetyl nitrate [$CH_3C(O)OONO_2$] or PAN, is noteworthy in the urban environment for a number of reasons. PAN is formed as a product of the photochemical oxidation of a range of VOCs, the simplest of which is acetaldehyde, CH_3CHO.

$$CH_3CHO + OH \rightarrow CH_3CO + H_2O \tag{20}$$

$$CH_3CO + O_2 + M \rightarrow CH_3C(O)OO + M \tag{21}$$

$$CH_3C(O)OO + NO_2 + M \leftrightarrow CH_3C(O)OONO_2 + M \tag{22}$$

PAN is much less soluble than HNO_3, so represents a NO_x reservoir rather than a permanent sink. The position of equilibrium for Reaction (22) is *highly* temperature dependent: the PAN decomposition rate increases by 17% per degree Kelvin at 288 K,[7] and the lifetime of PAN is a few months at upper tropospheric temperatures, but only an hour at 296 K. Consequently the formation of PAN followed by rapid cooling (*e.g.* through uplift of a polluted airmass), transport and subsequent descent/warming provides a mechanism by which NO_x can be exported to regions remote from the urban environment where the emissions originally occurred. Other alkyl nitrates, formed as a minor channel of the $RO_2 + NO$ reaction, can also (in principle) behave in a similar manner, but their formation rates are generally much slower than that of PAN. In the health context, PAN is a strong lachrymator, and is predominantly responsible for the eye-watering tendency of intense photochemical smog episodes. (The term "PANs" is sometimes applied to a whole range of peroxyacyl nitrate species, with a range of organic groups attached to the $-C(O)OONO_2$ functional group, and similar behaviour to peroxyacetyl nitrate.)

4.3 Night-Time Processes

At night, both OH production from $O(^1D) + H_2O$ chemistry and NO_2 photolysis cease, and the ozone production cycle is replaced by a chain of reactions converting NO and NO_2 to NO_3 and ultimately N_2O_5:

$$NO + O_3 \rightarrow NO_2 + O_2 \qquad (23)$$

$$NO_2 + O_3 \rightarrow NO_3 + O_2 \qquad (24)$$

$$NO_2 + NO_3 + M \leftrightarrow N_2O_5 + M \qquad (25)$$

N_2O_5 may be removed from the atmosphere through heterogeneous reaction with water vapour on the surface of atmospheric aerosol particles, forming nitric acid:

$$N_2O_5 + H_2O \rightarrow 2HNO_3 \quad \text{(surface)} \qquad (26)$$

The nitrate radical, NO_3, is the night-time analogue of OH: it acts as the primary oxidant, initiating the degradation of many hydrocarbons. NO_3 undergoes solar photolysis on a timescale of seconds during daylight, essentially switching off Reactions (25) and (26). NO_3 reactions with alkanes are generally too slow to be of significance; reactions with unsaturated and carbonyl compounds are much faster, and may comprise a significant night-time source of HO_x radicals. The equilibrium between NO_2, NO_3 and N_2O_5 is rapid (the lifetime of N_2O_5 with respect to thermal decomposition is about 1 minute at 298 K) thus NO_3 and N_2O_5 are tightly coupled in the atmosphere. In urban environments however, the rapid reaction between NO and NO_3 (27) usually prevents NO_3 and hence N_2O_5 building up, at least where fresh vehicle exhaust emissions (*i.e.* elevated levels of NO) are present.

$$NO + NO_3 \rightarrow 2NO_2 \qquad (27)$$

4.4 Timescales of Ozone Production

The timescale of the ozone production process can range from hours in very heavily polluted environments to a few days: under the meteorological conditions most commonly associated with ozone episodes over northwestern Europe and the UK, summertime anticyclonic periods in which airmasses have looped over Europe accumulating pollutants before arriving in the UK from the south-east quadrant, ozone is chemically formed along the airmass trajectory at rates of the order of 20–30 ppb per day (daytime chemical production, part of which is offset by night-time chemistry and deposition).[8] If a typical episode is characterised by ozone levels of the order of 90 ppb (a level at which significant health effects may be experienced by some individuals) compared with the rural background of around 40 ppb, several days are clearly required to achieve this level. In more extreme conditions, ozone production proceeds more rapidly: during the 2003 Western Europe heatwave event, a photochemical model constrained by *in situ* observations of long-lived species from suburban North-East London[9] determined the peak ozone production to be $17\,ppb\,h^{-1}$, with an average production rate during midsummer at midday of $7.2\,ppb\,h^{-1}$. Observations in Mexico City[10] have led to calculated ozone production rates of up to $50\,ppb\,h^{-1}$ during stagnant meteorological conditions.

The contrasting timescales between the $NO-NO_2-O_3$ photochemical steady state (minutes) and ozone production (hours to days), coupled with the predominance of NO over NO_2 in vehicle exhaust, leads to the evolution of the NO_x-O_3 mixtures from urban centres to rural areas: ozone levels are very low in most city centres, of the order of a few ppb at the roadside, as a consequence of the $NO+O_3$ reaction. As air is advected away from the urban centre, and is diluted with comparatively ozone-rich, NO_x-poor air from aloft, the $NO_2 : NO$ ratio increases and ozone levels increase; photochemical ozone production processes then lead to the highest ozone levels being encountered in suburban and rural regions tens to hundreds of km downwind of the urban conurbations.

4.5   Analysis of Ozone Sources

Due to the rapid interchange of O_3 and NO_2, it can be helpful to consider the total level of oxidant, O_x (O_3 and NO_2), and its variation with NO_x levels. Elevated levels of O_x may result from either regional ozone production episodes as outlined above (*i.e.* ozone rich air advected from elsewhere, into which local NO emissions occur converting O_3 to NO_2) or potentially from local sources. These include direct primary emissions of NO_2 (see comments below), intermolecular NO to NO_2 conversion *via* reaction (3), or local radical-driven NO to NO_2 conversion/ozone production. These regional and local contributions to O_x may be distinguished through their differing dependence upon NO_x levels – a plot of (the mixing ratios of) O_x as a function of NO_x determines the regional (NO_x-independent) contribution from the ordinate intercept, and the local (NO_x-dependent) contribution from the gradient (Clapp & Jenkin, 2001).[11] An example of this analysis applied to measurements made on the BT tower in central London during October/November 2007 is shown in Figure 3 – these data clearly show both a regional and local contribution to the total oxidant loading.

4.6   Limits on Ozone Production: NO_x-Limited and VOC-Limited

As both NO_x and VOCs are required for significant ozone production to occur, an understanding of the response of the atmosphere to changes in the abundance of either group of compounds is necessary to properly inform design of emissions control legislation aimed at reducing pollutant levels. At very low levels of NO_x, ozone is destroyed through photolysis; as NO_x levels increase

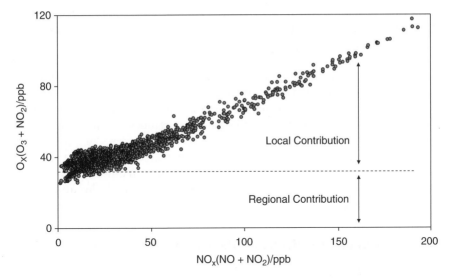

Figure 3 Dependence of oxidant O_x upon NO_x levels, indicating regional and local contributions to $(O_3 + NO_2)$. Data taken from the BT Tower, London during November 2007.

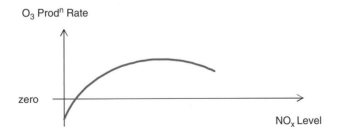

Figure 4 Simplified dependence of the ozone production rate upon NO_x level, for a given VOC abundance, sunlight *etc.*

ozone production increases through the propagation Reactions (16)–(18), until the $OH + NO_2$ chain termination process (19) becomes sufficient to compete with radical propagation and ozone production decreases once more, giving the form of the NO_x dependence shown schematically in Figure 4.

The precise form of the ozone production *vs.* NO_x curve depends upon the local chemical conditions (critically, the VOC loading) and upon environmental factors such as solar insolation. If NO_x levels are very high, as might be anticipated in urban centres, reducing NO_x levels will lead to increased ozone production (in that particular environment) and the system is termed VOC-limited. If NO_x levels are lower (commonly encountered as air moves out from the city centre), ozone production will not vary significantly with VOC levels, but will increase/decrease positively in response to changing NO_x

levels – the system is termed NO_x-limited. The transition from VOC-limited to NO_x-limited conditions as the airmass moves from city centre to suburban and rural environments arises in part from dilution, and in part as a consequence of the chemical lifetime of NO_x being less than that of the VOC cohort. A further consideration in emissions control legislation is that if NO_x emissions are reduced with the aim of limiting photochemical ozone production, the titration of ozone, Reaction (4), is also reduced and ozone levels close to the emission sources, *i.e.* in urban areas with high population densities, will increase towards the background level.

Determination of the precise regime a particular environment represents, and the associated formulation of air quality policy, requires a quantitative knowledge of both the chemical conditions/emissions and the atmospheric chemical processes responsible.

5   Modifications Particular to the Urban Environment

5.1   Radical Sources

While radical production in the background free troposphere is dominated by short-wavelength ozone photolysis and the $O(^1D) + H_2O$ reaction, in the urban environment many more sources are important (or even dominant), including: photolysis of HONO and carbonyl compounds, alkene ozonolysis and (at night) reactions of the nitrate radical with VOCs (here we distinguish between primary production of OH, HO_2 and RO_2 radicals from stable precursors, and secondary production/cycling processes, *i.e.* rapid NO-driven $RO_2 \rightarrow HO_2 \rightarrow$ OH interconversion).

Radical budgets have been constructed from speciated composition measurements for a limited number of urban locations. A common conclusion of such analyses is that primary production through ozone photolysis is only a minor radical production channel, with overall chain initiation driven by photolysis of nitrous acid (HONO), formaldehyde (HCHO), higher aldehydes (RCHO) or the reaction of ozone with alkenes:

$$HONO + h\nu \rightarrow OH + NO \tag{28}$$

$$HCHO + h\nu \rightarrow H + HCO \rightarrow 2HO_2 + CO \tag{29}$$

$$RCHO + h\nu \rightarrow H + RCO \rightarrow HO_2 + R'O_2 + CO \tag{30}$$

$$Alkene + O_3 \rightarrow OH + HO_2 + RO_2 + Products \tag{31}$$

The importance of non-primary [non $O(^1D) + H_2O$] sources of HO_x was illustrated by observations of OH and HO_2 radical concentrations performed at an urban background site in Birmingham (UK) during July 1999 and January 2000, *i.e.* summer and winter.[12] While levels of solar insolation were greatly reduced [mean local solar noon $j(O^1D)$ – the rate of photolysis of ozone to produce $O(^1D)$ atoms *via* reaction (8) – reduced by a factor of 15], OH levels only fell by a factor of approximately 2 (Figure 5). During this study, the primary sources of OH were found to be alkene ozonolysis (46% in summer, 62% in winter) followed by HONO photolysis (29 : 36%). The dominant radical (HO_x) sources were photolysis of HCHO and higher carbonyl species (68% of the total in summer, 24% in winter) and alkene ozonolysis (19% of the total in summer, 58% in winter).

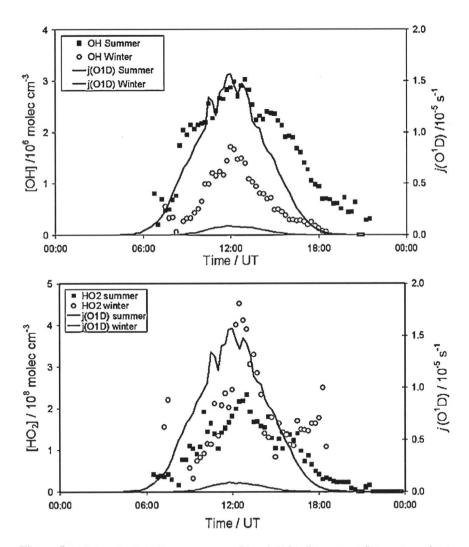

Figure 5 Mean diurnal OH (upper panel) and HO_2 (lower panel) concentrations, and photolysis rate for $O_3 \rightarrow O(^1D) + O_2$ (upper line: summer; lower line: winter), measured in Birmingham, UK, during summer 1999 and winter 2000. From Heard *et al.*[12]

Nitrous acid (HONO) is a recognised pollutant of the urban environment (and elsewhere). It is formed by the recombination of OH radicals with NO in the gas phase, and by a number of heterogeneous mechanisms. Studies have shown that HONO concentrations in urban environments increase during the night-time, reaching levels of up to several ppb, before rapidly falling after sunrise due to long-wavelength solar photolysis releasing OH and NO; the early morning release of OH radicals can act to start up the daytime chemistry before other sources of OH become important later in the day. There is uncertainty regarding the nature of this night-time HONO source, but it is thought to arise from heterogeneous interactions involving disproportionation of NO_2 with surface water, and possibly reaction of NO_2 with surface-adsorbed reductants.[13]

Recent developments in HONO measurement technology have lowered instrumental detection limits, and resulted in observations of smaller, but still unexpectedly high, concentrations of HONO in daytime, in the urban atmosphere and elsewhere (*e.g.* Kleffmann *et al.*, 2003)[14] with levels above 200 ppt observed. Given the rapid solar photolysis of HONO, such levels imply a substantial daytime HONO source, over and above recombination of OH with NO, must also exist. The source of this daytime HONO is uncertain but the reported concentrations indicate that it must be more significant than the night-time source – *i.e.* it is photoenhanced. Recent laboratory studies have identified potential new sources of HONO, reactions of NO_2 upon humic acids ubiquitous to surfaces in the urban environment which could explain these observations.[15] Budget estimates suggest that this mechanism – surface HONO production – may in fact be the dominant radical source in the lowermost atmosphere, if NO_x levels and abundance of the adsorbed co-reactants are sufficient – conditions likely to be fulfilled in the urban boundary layer.

5.2 VOC Speciation and Abundance

The chemical complexity of the urban atmosphere, in particular the VOC loading, is a major challenge to attempts to model urban atmospheric chemistry: OH concentrations, and consequently RO_2 production rates, depend upon inclusion of all significant co-reactants; given the complexity of (for example) gasoline composition, further modified by combustion, comprehensive measurement of all VOC species in the atmosphere is very hard to achieve. In the free troposphere, the VOC OH sink is dominated by CH_4 and CO, while in urban areas larger hydrocarbons dominate – for example during the TORCH campaign to the North-East of London, methane only accounted for 7–13% of the total (measured) OH VOC sink.[16] Conventional approaches to measurement of VOC using chromatographic methods are able to detect most small molecules, such as C_2–C_6 alkanes and alkenes, and some aromatic species such as benzene and toluene; however evidence suggests that a substantial additional gas-phase VOC pool also exists. "Comprehensive" chromatography, employing sequential separations based upon volatility and polarity

coupled with mass-spectroscopic analysis, have revealed many more compounds to be present than was previously thought: measurements taken in
Melbourne city centre using this method revealed at least 500 compounds to be
present, including over 100 previously unobserved multisubstituted monoaro-
matic and oxygenated VOCs, with only 20% of the total reactive NMHC mass
lying within the C_2–C_6 range.[17] The presence of many more large volatile VOCs
than previously measured in the urban environment implies that models
may underestimate the production of ozone and secondary organic aerosol,
although this effect may be masked by the lumping approaches necessarily
adopted in complex chemical schemes.

5.3 Primary Emissions of NO_2

In the recent past, it has commonly been assumed that the overall ratio of NO :
NO_2 emitted from the motor vehicle fleet has a value of approximately 0.05 :
0.95. It is recognised that this is a considerable simplification, with different
engine technologies displaying different emissions ratios under different driving
cycles; in particular diesel engines tend to emit a higher fraction of NO_2.

Measurements of NO_x in the urban and suburban atmosphere of the UK
have shown a decrease since the mid-1990s, as would be anticipated with
emissions control legislation coming into force and the introduction of new
vehicle technologies. Urban NO_2 levels however have not displayed as large a
decrease; while the NO_2 : NO_x ratio would be expected to increase at lower
NO_x levels, the upward trend in NO_2 exceeds that expected from NO_x photochemistry.[18] Possible explanations for this observation include increased
primary emissions of NO_2 (potentially associated with greater usage of diesel
vehicles, or an increase in NO_2 emissions as a consequence of control technologies such as catalytically regenerative particle traps fitted to buses), an
increase in the hemispheric background ozone concentration (leading to
increased NO to NO_2 conversion), or emissions of other species which promote
the formation of (or are measured as) NO_2, such as HONO.

In the case of London, the AQEG study found that the elevated NO_2 could
be explained by a combination of increased usage of diesel vehicles and/or
increased NO_2 emissions as a consequence of new exhaust technologies. The
fraction of NO_x emitted as NO_2 (f-NO_2) was found to range from 20–70% for
vehicles compliant with the Euro-III emission standard (a range some way from
the 5% f-NO_2 previously adopted), while the introduction of regenerative
particulate traps for buses, in which NO_2 levels are deliberately elevated to
facilitate combustion of soot and so regenerate the trap, also leads to greater
NO_2 emissions. The primary emission fraction of HONO from vehicle exhaust
is not well known (see comments above), but could also contribute to this trend
if it is increasing with changing vehicle technology: Jenkin *et al.*[19] calculated
that during pollution episodes, increases in the fraction of NO_x emitted as
HONO over the 0–5% range had about five times the effect upon oxidant as the
equivalent increases in f-NO_2, as a consequence of the increased (radical driven)

NO-to-NO_2 oxidation, and that the increased radical and NO_2 levels increased the rate of nitric acid formation, and hence levels of particulate nitrate, close to the emission region.

6   Particulate Phase Chemistry

Particulate matter is an important component of the atmospheric chemical system, in particular with relation to health effects in the populated, urban environment. Particulate matter is released to the atmosphere as primary material, from sources such as incomplete combustion (soot/smoke), dust, sea salt and brake pad/tyre wear, and is formed as secondary material from the condensation of low volatility gases such as sulfuric acid, larger organic compounds and the co-condensation of high-temperature combustion products in vehicle exhausts. The term aerosol is commonly used to refer to atmospheric particulate matter, although strictly the term refers to both the condensed phase material and the gas it is suspended within.

Particulate matter is commonly characterised by its size, more precisely by aerodynamic diameter (amongst other metrics; as atmospheric particles exhibit a vast range of morphology the concept of aerodynamic diameter is commonly used to assess a particle size – a particle with an aerodynamic diameter of 1 micron will exhibit the same inertial properties as a sphere with a diameter of 1 micron and a density of 1 g cm^{-3} – irrespective of the actual size, shape or density of the particle). A tri-modal size distribution is commonly observed in the lower atmosphere: nucleation or Aitken mode, below approximately 0.1 μm in diameter; accumulation mode (0.1–2.5 μm), and the coarse mode (>2.5 μm diameter). Secondary particles are formed from the condensation of low-volatility gases in the nucleation mode (or, more commonly, material condenses onto existing particles). Particles grow in the accumulation mode through coagulation and condensation. Primary particles, generated mechanically, are mostly emitted in the coarse mode. The lifetime of accumulation mode particles is largely determined by frequency of precipitation, while nucleation mode particles can more readily diffuse and undergo agglomeration, and coarse mode particles undergo gravitational settling at appreciable rates. The greatest numbers of particles occur in the nucleation mode, the greatest mass in the coarse mode (with an appreciable contribution from the accumulation mode). The smallest particles, while not dominating the total mass, are increasingly thought to have the greatest potential health effects as they can penetrate most deeply into the alveolar structure. It is worth noting that these particle size classifications are not fixed, and various definitions will be encountered in different resources, particularly with regard to the *fine* and *ultrafine* fractions (generally below 2.5 and 0.1 μm, respectively).

Legislative air quality objectives are currently defined in terms of particle mass concentration; at the time of writing (late 2008) the United Kingdom Air Quality Objective is for PM_{10}, that fraction of particles with an aerodynamic diameter below 10 μm, although recently introduced European legislation has required limits on $PM_{2.5}$ – it is increasingly becoming evident that the fine fraction of aerosol has the greatest adverse impacts upon human health. It is instructive to consider the amount of condensed phase material over a typical (western) city: taking the example of London, a mean mass concentration for PM_{10} of the order of $20\,\mu g\,m^{-3}$ is commonly observed. Considering the area of central London (*ca.* $400\,km^2$) and assuming a 1 km boundary layer height, an aerosol loading of 8 tonnes is obtained.

6.1 Chemical Composition

The principal components of urban aerosol particles are inorganic chemical constituents such as sulfate (SO_4^{2-}), nitrate (NO_3^-), chloride (Cl^-) and ammonium (NH_4^+), organic compounds, elemental carbon (soot, a composite of black carbon and low volatility organics formed as exhaust gases cool) and trace quantities of metals such as lead, arsenic and vanadium (found in fuels). Sulfate and nitrate originate in sulfuric and nitric acid, which in turn are derived from the gas-phase oxidation of sulfur–SO_2 and NO_x, respectively. Ammonia, of primarily biological origin, readily dissolves in aqueous particles and neutralises the sulfate and nitrate, which are usually found as ammonium sulfate [$(NH_4)_2SO_4$] or ammonium nitrate (NH_4NO_3) in urban regions. Measurement of the total aerosol composition is challenging due to instrumental limitations and the volatile nature of some components; typical overall aerosol particle composition deduced from combined measurements and mass closure procedures for PM_{10} and $PM_{2.5}$ in an urban location are shown in Figure 6 (taken from Yin & Harrison).[20]

6.2 Secondary Particle Formation

Combustion of sulfur-containing fuels (coal, diesel, gasoline) leads to the release of sulfur dioxide (SO_2) to the atmosphere, where it is readily oxidised to sulfuric acid (H_2SO_4). Sulfuric acid has a low vapour pressure in combination with water, and condenses under atmospheric conditions to form aqueous sulfate particles. The particle formation process is enhanced in the presence of species such as ammonia (NH_3), which co-condense with H_2SO_4 and water vapour to form a nucleation-mode aerosol. Nascent ultrafine aerosol particles grow through condensation of semi-volatile gases and coagulation leading to the complex composition observed in the ambient atmosphere.

The oxidation of volatile organic compounds (VOCs) by oxidants such as OH, ozone and the nitrate radical leads to the production of functionalised species, such as aldehydes and ketones which have a carbonyl unit, C=O, incorporated into the molecular structure. The presence of functional groups increases the attractive forces between such molecules, reducing their vapour pressure. As vapour pressure is also dependent upon molecular size, if the initial VOC is sufficiently large (bigger than around 5–7 carbon atoms depending upon structure), its oxidation products may enter the condensed

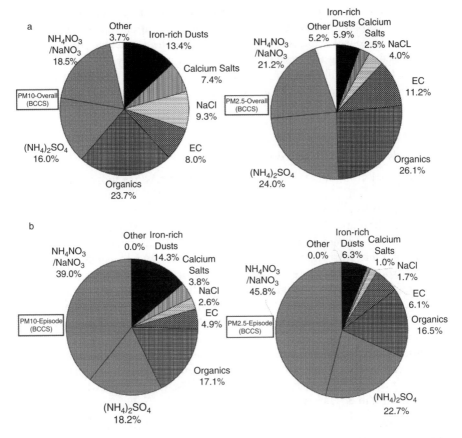

Figure 6 Composition of $PM_{2.5}$ and PM_{10} in an urban location (city centre, Birmingham, UK) deduced from application of a mass closure model to measured chemical composition.[20] Notes: EC = elemental carbon; BCCS = Birmingham City Centre Site; upper panels (a) are averages of all measured data, lower panels (b) correspond to pollution episode days where PM_{10} exceeded $50\,\mu\mathrm{g}\,\mathrm{m}^{-3}$.

phase and contribute to aerosol growth. In the urban environment, the oxidation of aromatic molecules (those based upon a benzene ring structural unit such as benzene, toluene and xylene) leads to the incorporation of substantial quantities of such secondary organic aerosol (SOA) in the fine particle mode. More widely, biogenic VOCs, particularly monoterpenes ($C_{10}H_{16}$ species such as α-pinene and limonene) are also major SOA precursors.

The composition of aerosol particles changes with time in the atmosphere through chemical effects in addition to physical processes such as condensation and evaporation. Chemical processing, both internally and driven by external reagents, notably gas-phase oxidants such as OH radicals and ozone, may change particle physical characteristics (size, volatility) and chemical composition; this in turn may affect key characteristics such as particle lifetime, heterogeneous reactivity, interaction with solar radiation and potential to act as cloud condensation nuclei. Our understanding of the chemical evolution of aerosol is in its infancy, but some measurements indicate that the chemical lifetime of organic constituents in urban aerosol is comparable to, or shorter than, the physical lifetime of the particles.[21]

6.3 Other Impacts of Aerosol

Aerosol particles provide a reaction surface for the removal (and production) of gaseous species. Two processes of significance for the urban environment relate to the removal of NO_x from the gas phase through the heterogeneous hydrolysis of N_2O_5 [see Reaction (26)], and gas vapour equilibria such as that between the gases ammonia and nitric acid, and solid-phase ammonium nitrate, which limits the concentrations of these species which are found in the gas phase [the levels are determined or buffered by the equilibrium position for reaction (32), rather than gas phase production and loss processes].

$$N_2O_5(g) + H_2O(l) \rightarrow (surface) \rightarrow 2HNO_3(aq) \tag{26}$$

$$NH_3(g) + HNO_3(g) \leftrightarrow NH_4NO_3(s) \tag{32}$$

Aerosol particles affect the transfer of radiation through the atmosphere, through absorption and scattering (much of the purple-brown haze associated with "smog" events over cities arises from the attenuation of visible radiation by particulate matter), with wider impacts upon climate: scattering of solar radiation by anthropogenically produced sulfate aerosols is thought to have been at least partly responsible for offsetting the greenhouse-gas induced radiative forcing during the third quarter of the twentieth century ("global dimming").[22] Particles also have a strong effect upon global climate through their role as cloud condensation nuclei, directly affecting the propensity of clouds to form and through indirect effects of the induced droplet size distribution affecting cloud brightness and lifetime.

7 Modelling Urban Atmospheric Chemistry

A comprehensive discussion of approaches to the simulation of chemical processes within city environments is beyond the scope of this article. Accurate representation of urban environments within atmospheric models poses a major challenge due to the chemical and dynamic complexity and high spatial variability. Chemistry-only schemes, such as zero-dimensional box models and Lagrangian trajectory models, can afford (computationally) to implement detailed chemical mechanisms such as the SAPRC Mechanism,[23] the near-explicit Master Chemical Mechanism (MCM; http://mcm.leeds.ac.uk)[24] or lumped Regional Atmospheric Chemistry Mechanism,[25] while as dynamical detail is incorporated, the chemical schemes must inevitably be simplified.

Complex chemical mechanisms are used to inform development of strategies for emissions control legislation, to (for example) determine the ozone production efficiency for different VOC species (*e.g.* Derwent *et al.*).[26] Such mechanisms draw upon laboratory studies of individual chemical reactions for their fundamental kinetic and photochemical parameters, however even in the case of near-explicit schemes such as the MCM, many details of the chemical processes are not well characterised (for example, the daytime sources of HONO, radical yields from alkene–ozone chemistry, true VOC populations) and some fundamental details of the mechanism and kinetics are still the subject of disagreement (for example, the rate constant for the critical $OH + NO_2$ reaction)[27] or are simply unmeasured for larger and more complex hydrocarbons. Techniques such as structure–activity relationships and automated mechanism development are useful in such cases. The overall mechanisms may then be tested against (or tuned using) simulation chamber studies of VOC oxidation. If the model reproduces the observed VOC, NO_x and ozone evolution one has some confidence that the chemical mechanism may be accurate. Such models may then be used to inform policy development – for example, ranking different hydrocarbons by their ozone production potential. A full simulation of the urban atmospheric environment requires consideration of the detailed chemical interactions in both the gas and condensed phases, spatially resolved emissions inventory, on- or off-line dynamical processes (turbulence, mixing, entrainment *etc.*), together with the surrounding atmospheric boundary conditions within which the model can run. It is perhaps unlikely that such a model can reasonably be constructed to represent a given city, and approaches such as measurement of total reactivity offer a way forward by simplifying some of the atmospheric complexity.

8   Conclusions

The gas-phase chemistry of contemporary urban atmospheric environments is dominated by transport emissions of NO_x and VOCs, the NO_x–ozone steady state and the oxidation of VOCs leading to peroxy radical mediated NO to NO_2 conversion and ozone production. By-products of VOC oxidation, such as PAN, are substantial pollutants on the local scale and potentially lead to the wider export of urban emissions. Particulate matter is a key component of the urban chemical mix, with major health and environmental impacts in its own right, and which interacts bi-directionally with the gas-phase constituents.

References

1. *World Urbanization Prospects: The 2007 Revision*, United Nations: United Nations Department of Economic and Social Affairs/Population Division, 2008.
2. P. Brimblecombe, *The Big Smoke*, Methuen, London, 1987.

3. Y. Li, J. Thomas, J. Jackson, K. King, T. P. Murrells, N. Passant and G. Thistlethwaite, *Air Quality Pollutant Inventories for England, Scotland, Wales and Northern Ireland*, AEA Report AEAT/ENV/R/2678, 2008.

4. M. Z Jacobson, *Environ. Sci. Technol.*, 2007, **41**, 4150.

5. J. S. Bower, G. F. J. Broughton, J. R. Stedman and M. L. Williams, *Atmos. Environ.*, 1994, **28**, 461.

6. P. A. Leighton, *Photochemistry of Air Pollution*, Academic Press, New York, 1961.

7. S. P. Sander, R. R. Friedl, D. M. Golden, M. J. Kurylo, G. K. Moortgat, H. Keller-Rudek, P. H. Wine, A. R. Ravishankara, C. E. Kolb, M. J. Molina, B. J. Finlayson-Pitts, R. E. Huie and V. L. Orkin, *Chemical Kinetics and Photochemical Data for Use in Atmospheric Studies, Evaluation No. 15*. JPL Publication 06-2, NASA Jet Propulsion Laboratory, Pasadena, CA, 2006.

8. M. E. Jenkin, T. J. Davies and J. R. Stedman, *Atmos. Environ.*, 2002, **36**, 999.

9. K. M. Emmerson, N. Carslaw, D. C. Carslaw, J. D. Lee, G. McFiggans, W. J. Bloss, T. Gravestock, D. E. Heard, J. Hopkins, T. Ingham, M. J. Pilling, S. C. Smith, M. Jacob and P. S. Monks, *Atmos. Chem. Phys.*, 2007, **7**, 167.

10. E. C. Wood, S. C. Herndon, T. B. Onasch, J. H. Kroll, M. R. Canagaratna, C. E. Kolb, D. R. Worsnop, J. A. Neuman, R. Seila, M. Zavala and W. B. Knighton, *Atmos. Chem. Phys.*, 2009, **9**, 2499.

11. L. J. Clapp and M. E. Jenkin, *Atmos. Environ.*, 2001, **35**, 6391.

12. D. E. Heard, L. J. Carpenter, D. J. Creasey, J. R. Hopkins, J. D. Lee, A. C. Lewis, M. J. Pilling, P. W. Seakins, N. Carslaw and K. M. Emmerson, *Geophys. Res. Lett.*, 2004, doi:10.1029/2004GL020544.

13. G. Lammel and J. N. Cape, *Chem. Soc. Rev.*, 1996, 361.

14. J. Kleffman, R. Kurtenbach, J. Lörzer, P. Wiesen, N. Kalthoff, B. Vogel and H. Vogel, *Atmos. Environ.*, 2003, **37**, 2949.

15. K. Stemmler, M. Amman, C. Donders, J. Kleffman and C. George, *Nature*, 2006, **440**, 195.

16. J. D. Lee, *et al.*, *Atmos. Environ.*, 2006, **40**, 7598.

17. A. C. Lewis, N. Carslaw, P. J. Marriott, R. M. Kinghorn, P. Morrison, A. L. Lee, K. D. Bartle and M. J. Pilling, *Nature*, 2000, **405**, 778.

18. *Air Quality Expert Group: Trends in Primary Nitrogen Dioxide in the UK*, DEFRA Publications, 2007.

19. M. E. Jenkin, S. R. Utembe and R. G. Derwent, *Atmos. Environ.*, 2008, **42**, 323.

20. J. Yin and R. M. Harrison, *Atmos. Environ.*, 2008, **42**, 980.

21. A. L. Robinson, N. M. Donahue and W. F. Rogge, *J. Geophys. Res.* 2006, doi:10.1029/2005JD006265.

22. *IPCC, Climate Change 2007: The Physical Science Basis. Contribution of Working Group I to the Fourth Assessment Report of the Intergovernmental Panel on Climate Change*, ed. S. Solomon, D. Qin, M. Manning, Z. Chen, M. Marquis, K. B. Averyt, M. Tignor and H. L. Miller, Cambridge University Press, 2007.

23. W. P. L. Carter, *Atmos. Environ.*, 1990, **24**, 481.
24. M. E. Jenkin, S. M. Sauders and M. J. Pilling, *Atmos. Environ.*, 1997, **31**, 81.
25. W. R. Stockwell, F. Kirchner, M. Kuhn and S. Seefeld, *J. Geophys. Res.*, 1997, **102**, 25847.
26. R. G. Derwent, M. E. Jenkin, N. R. Passant and M. J. Pilling, *Atmos. Environ.*, 2007, **41**, 2570.
27. S. Valluvadasan, D. B. Milligan, W. J. Bloss and S. P. Sander, *Proceedings of the 19th International Symposium on Gas Kinetics*, Orleans, July 2006.

Air Pollution in Underground Railway Systems

IMRE SALMA

ABSTRACT

Metropolitan underground railway transport systems are a very important part of the urban traffic since they carry millions of passengers per day in a number of cities around the world. They represent a segregated traffic microenvironment because of their closed character and restricted ventilation, lack of sunlight, specific emission sources and meteorological conditions. Concentrations of some air pollutants including aerosol particles are usually higher in underground railways than in corresponding external outdoor areas, which can considerably increase commuters' daily exposures. At the same time, the chemical composition and properties of indoor aerosol particles differ substantially from those for outdoor air; the particles are larger and heavier, are mainly composed of Fe, and contain less soot, and, therefore, their impact on morbidity and mortality is expected to be different. Nevertheless, some transition metals, *i.e.*, Fe, Mn, Ni, Cu and Cr can be of concern. In the present paper we overview the physical and chemical properties and behaviour of the aerosol particles, as well as their major emission sources, compare the information and knowledge available for the underground railways, and discuss their health implications and conclusions in general.

1 Introduction

It is estimated that people living in urban areas in developed countries spend approximately 8% of their daily time commuting. Concentrations of pollutants

Issues in Environmental Science and Technology, 28
Air Quality in Urban Environments
Edited by R.E. Hester and R.M. Harrison
© Royal Society of Chemistry 2009
Published by the Royal Society of Chemistry, www.rsc.org

in traffic microenvironments are higher than in other urban and urban background environments. Of them, metropolitan underground railway transport systems (metros, subways or tubes) represent a segregated microenvironment because of their closed character and restricted ventilation, lack of sunlight (photochemistry), specific emission sources and meteorological conditions.

Underground railway systems are urban, electric transport systems with high capacity and a high frequency of service mainly realised in tunnels separated from the other traffic. The first steps of their operation started in London (1863), Chicago (1892) and Budapest (1896). The first modern underground railway line was built in London, and it was opened in 1890. Since then, some 116 cities in Europe, America, Asia and North Africa opened their metro systems. Subways became a very important part of the urban traffic. Metro networks carried some 155 million passengers per day in 2006. The most frequent average time spent on one journey is between 10 and 20 min, which adds up to 1–1.5 h that an ordinary commuter spends in the system a day.[1]

There are important differences among underground railway systems around the world, which include the differences in the length of the lines and number of stations as well as in construction and operational circumstances, engineering systems, number of train passages, passenger densities, ventilation and air conditioning systems for stations and vehicles, dimensions of underground spaces, cleaning frequencies and other factors.[2–27,39] In most subways, the trains run on steel wheels while some or all vehicles in Hong Kong, Montreal and Mexico City are covered by rubber lining.[11,21,24,28] Braking systems applied are also diverse; they consist of either pneumatic friction brakes (asbestos-free composite brake blocks) or electric brakes complemented by pneumatic braking at low velocities. Emergency braking systems involve friction brakes with or without spraying sand on the rails.[28] In Rome, approximately 400 tons of silica sand is used by the underground railway lines each year.[20] All these circumstances largely influence the air pollution levels in subways. In addition, the air quality in metros is also affected by the ambient air pollution above the stations, which changes from network to network, from line to line, and it also varies in time and space within a line and station.[28] These features complicate the comprehensive characterisation and comparison of underground railways.

There is an increasing number of studies which have considered the air quality in subways. Most of them dealt with the particulate matter (PM)[2–27,39] because it is this air pollutant that is abundant, that has important sources and can have health effects for the public in metros. Particles with an aerodynamic diameter (AD) $\leq 10\,\mu m$ (PM_{10} size fraction) can penetrate into the human lungs. Particles with an $AD \leq 2.5\,\mu m$ ($PM_{2.5}$ size fraction) are recognised because of their potentially harmful chemical composition, their larger surface area, and their larger atmospheric residence time. These size fractions are therefore accorded particular attention and interest. Other investigations have included some criteria pollutant gases,[12,23,24] polycyclic aromatic hydrocarbons (PAHs),[2,9] volatile organic compounds[24,27] or microorganisms.[10,13] The measuring approaches, methods, techniques and instruments applied in the aerosol studies are at least as diverse as the subways themselves. Overview or discussion of the experimental procedures, calibration and correction methods are beyond the scope of the present summary due to length limitations. We just note that the experiments include the collection of various size fractions of aerosol particles by filtration or impaction on different substrates for subsequent chemical analyses or other investigations, and *in situ* measurements that are based on very different (electrodynamical, optical, mechanical or other) principles and realisations, while others use personal exposure equipment. Considering these, it is understandable that the measured results derived can not always be linked or compared directly, since they are not fully representative and of the same quality and reliability. Nevertheless, we will attempt – relying on these studies – to provide an overview and summarise the relevant aerosol properties, to identify their tendencies that can be generalised for most or many underground railways, and to highlight some peculiarities of the particulate air pollution in underground railway systems.

2 Physical and Chemical Properties

2.1 Particle Mass and Number Concentrations

Particle mass and number concentrations have been measured in a number of underground railways. The studies show that (time averaged) mean PM_{10} mass concentrations vary substantially within an underground area.[18,25] Platforms have the highest levels of mass concentration, while concentrations in passenger carriages and station precincts (vestibule and corridors), respectively, are generally smaller. The increase in mean concentrations from precincts to platforms can be very marked (up to 40–50%) in some metros. In air-conditioned subway vehicles, mean concentrations are generally smaller by 15–20% than the levels for platforms, suggesting that filtration provided by air conditioners is an effective way in reducing (coarse) mass concentrations. Driver compartments (cabs) belong to the most polluted areas within the workers' activity spaces. They exhibit mean concentrations that are comparable with platforms. Somewhat less spatial variability is observed for the $PM_{2.5}$ size fraction; platforms, passenger carriages and driver compartments often show similar mean concentrations.

Typical ranges and averages of the PM mass available for various underground railways are summarised in Table 1. It is seen that there are substantial differences in the data. This could be partly explained by the differences in the methodological approaches and experimental circumstances, such as, in the measuring locations actually selected, stationary or mobile types of the measurements, season, duration and timing of the measurements, or in the boundary conditions for averaging (see the footnote of Table 1). Therefore, these values can not be used directly for categorising or rating the subway systems. The explanation also involves the differences among the various subways mentioned in Section 1. The mass concentrations presented are greater, usually several-fold, than for the outdoor areas just above or near the metro lines. This is demonstrated by the fact that mass concentrations in

Table 1 Typical range and average particulate mass concentration in $\mu g\ m^{-3}$ for different size fractions in various metropolitan underground railways.

City	PM_{10} size fraction		$PM_{2.5}$ size fraction		Reference
	Range	Average	Range	Average	
Beijing	–	325[a]	–	113[a]	23
Berlin	141–153	147[b]	–	–	2
Budapest	25–232	155[c]	–	51[d]	22
Cairo	794–1096	983[e]	–	–	10
Guangzhou	26–123	67[f]	–	44[f]	12
Helsinki	–	–	23–103	47–60[g]	18
Hong Kong	23–85	44[h]	21–48	33[h]	11
London	500–1120	795[i]	–	246[j]	3–7
–	1100–1500	–	270–480	–	8
Mexico City	–	–	8–68	39[k]	24
New York City	–	–	–	62[l]	13
Paris	–	320[m]	–	93[m]	39
Prague	10–210	103[n]	–	–	19
Rome	71–877	407[o]	–	–	20
Seoul	238–480	359[p]	82–176	129[p]	26,27
Stockholm	212–722	468[q]	105–388	258[q]	14
–	105–388	357[r]	220–440	288[s]	15
Taipei	11–137	32–66[t]	7–100	27–44[t]	25
Tokyo	30–120	72[u]	–	–	9

[a] Average mass concentrations within trains.
[b] Overall average PM_{10} mass data from 07:00 h over 9 h inside trains.
[c] Average of 30 s mean PM_{10} mass data for working hours.
[d] Average for $PM_{2.0}$ mass from 06:00 h over 12 h for workdays.
[e] Average PM_{35} mass for 1 h between 10:00 and 18:00 h near a ticket office.
[f] Overall average of personal exposures for 2.5 h in an underground rail line traversing the inner city.
[g] Average hourly mean $PM_{2.5}$ mass data at different stations from 06:00 h over 12 h for workdays.
[h] Personal exposure for 25–30 min in underground railway system.
[i] Exposure to PM_9 mass for 3 h inside trains for various lines.
[j] 8 h personal exposure to $PM_{2.5}$ mass with time inside trains and on platforms.
[k] Overall average of personal exposures to PM2.5 size fraction for *ca.* 1 h.
[l] Exposure integrating *ca.* 5 h in underground railway stations and 3 h riding in trains.
[m] Average mass concentration on a platform for 7:00–10:00 and 15:00–21:00 on several weekdays.
[n] Mean personal exposure to PM_{10} mass for 11 min intervals spent in underground spaces of metro stations over *ca.* 1 year.
[o] Overall average PM_{10} mass data for five different platforms on two lines.
[p] Overall geometric mean concentration on platforms at various stations for 24 h.
[q] Average hourly mean mass data from 07:00 h over 12 h for workdays.
[r] Mean PM_{10} mass for workdays.
[s] Mean for *ca.* $PM_{2.0}$ mass data over 3 h in the afternoon.
[t] Average concentrations on platforms at various stations for 1 h randomly distributed.
[u] Average concentrations on platforms at various stations for *ca.* 3 h during Sundays.

vehicles increases when the trains enter the tunnel and it decreases when they come out.[18] There are some polluted cities where the ordinary ambient mass concentrations are so large that the metro networks have lower levels than the outdoor urban environments.[11,12,24] The effect of washing on indoor aerosol in the Stockholm metro was studied and found to be of limited effectiveness, reducing the PM_{10} mass by approximately only 13%.[14] Measurements in the Prague underground railways were made immediately after a complete clean up and reconstruction after severe floods in the city in August 2002, which most likely explains the low PM_{10} levels found.[19] The conclusions of the Stockholm and Prague studies are not contradictory, because the washing in Stockholm was confined to the walls and tracks between the platforms, and was thus less thorough.

The concentration data in Table 1 can be related to the air quality by comparing them to health limits. EU or US EPA 24 h PM_{10} ambient limits and US EPA 24 h $PM_{2.5}$ ambient limit for the public were chosen as guidelines for such comparisons[30,31] even though the ambient limits do not apply indoors or in tunnels. In some countries, there are PM_{10} limit values for indoor environments as well, which are virtually equal to the corresponding ambient limits.[26] It is evident that almost all underground railway systems exhibit greater (sometimes substantially greater) concentrations than the daily ambient EU PM_{10} limit which is $50 \,\mu g \, m^{-3}$. It has to be mentioned here that this limit is frequently exceeded in many areas in Europe, especially during temperature inversions in winter. As far as the US EPA limits are concerned, the daily ambient PM_{10} and $PM_{2.5}$ limit values are 150 and $65 \,\mu g \, m^{-3}$, respectively; thus, concentrations for a number of subways are already below these limits.

Number concentrations of aerosol particles that have an $AD > 0.5 \,\mu m$ usually follow the tendencies in the particulate mass and vary substantially.[13] The number of smaller particles is much more uniform within the underground spaces.[18] Travelling does not affect the number concentrations of these particles. Generally, the number count of particles in metros is smaller than outdoors above the stations due to the infiltration of outdoor particles with losses both during infiltration and by deposition within the underground spaces.[3] The number concentrations in metros reflect the above-ground number concentrations near the stations, and their typical value ranges from 10^4 to $10^5 \, cm^{-3}$ (refs. 3,13,18). However, intermittent and local increases of ultrafine particles in occupied underground spaces cannot be excluded.

2.2 Time Variation

Particulate mass (especially for the $PM_{2.5}$ size fraction) and number concentration for the ultrafine particles in metros are influenced by the corresponding ambient concentrations above the ground.[18] Time variation for these quantities follows a seasonal pattern[9,13] which is often observed for the ambient air. In addition, micrometeorology has an effect on the concentrations in subways because it affects both the ventilation of underground systems and the ambient air concentrations, and, therefore, some day-to-day fluctuations in the mean particle mass and number concentrations are also observed.[5] Average concentrations during weekends or holidays are smaller due to the lower intensity of the main emission sources (see Section 3).

The particulate mass concentration on platforms also varies on both diurnal and shorter (tens of seconds) time scales. These changes are rather remarkable, consistent from one day to the next day, and show some periodicity. Figure 1 demonstrates the temporal variation of the PM_{10} mass and wind speed (WS) for 24 h as 30 s measured values and as 1 h smoothed curves in the Budapest metro as an example.[22] The shape of the diurnal variation is driven by the vehicle frequency which explains the daily cycle, and it is comparable to that for typical kerbside areas since the time-activity patterns are common.

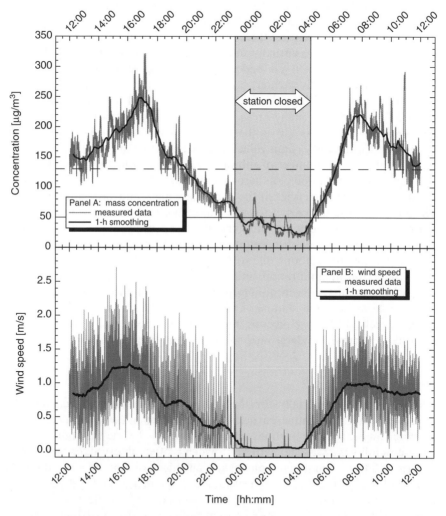

Figure 1 Temporal variation of PM_{10} mass concentration (A) and wind speed (B) as measured 30 s averages and 1 h smoothed curves. The ambient daily EU health limit of $50\,\mu g\,m^{-3}$ for PM_{10} mass concentration, and the 24 h mean mass concentration of $130\,\mu g\,m^{-3}$ are also shown in A by solid and dashed lines, respectively.[22] (Reproduced with kind permission of Elsevier.)

Both the PM_{10} and WS curve show two maxima (one between 07:30 and 08:00 h and the other around 17:00 h) corresponding to the morning and afternoon rush hours, when the train frequency is the greatest. Between the rush hours, the curves show a local minimum (at approximately 12:00 h). For some metros, this minimum is broader (as a plateau) or can be less evident.[13,18] Particle mass decreases rapidly after the departure of the last train (at approximately 23:20 h) which indicates a rather short residence time for aerosol particles in the air. This means that coarse particles dominate in the mass size distributions. Some abrupt changes in the data that occur while the station was closed are attributed to cleaning of the station (*e.g.* the peak in the PM_{10} mass at approximately 00:30 h) or they are explained by the motion of diesel-driven work cars involved in overnight freight shipment, maintenance and service activities (peaks in both PM_{10} and WS at approximately 02:00 h and just before 03:00 h). During weekends, mass concentrations generally start to rise with the train traffic, they reach a maximum in the morning hours and then maintain a fairly steady level until the evening hours after which they decline monotonically to a minimum at night.[14] In other investigations, no repeatable diurnal pattern in the mass concentrations is observed most likely due to the actual realisation of the rush hours (the number of vehicles in trains is increased, and the train frequency is increased to a lesser extent).[18]

Figure 2 displays wind direction (WD), WS, and PM_{10} mass data for a selected time interval (during which the coincident effect of trains running in the opposite direction in the other bore was well separated).[22] Changes in WS and WD reflect well the vehicle motion. A broad maximum for the WS curve is observed, corresponding to the arrival of trains in the station, standing at the platform, and departure from the station. The sudden change in WD and decreasing tendency of the WS curve indicate the end of the departure of trains from the station. Two maxima are observed on the PM_{10} mass curve for each arrival-standing-departure cycle. The first maximum is greater than the second, and occurs during the arrival phase. The increase in PM_{10} mass was caused by vehicles pushing in polluted air from the bore (by piston effect) as they entered the station, and PM_{10} mass decreases during deceleration. During passenger entry and exit, the airflow from the bore is maintained (as indicated by the WS curve), which results—together with motion of the commuters—in another increase in PM_{10} mass. Train departure pulls in some polluted air from the bore, as well as cleaner air from the vestibule and corridor, and therefore PM_{10} mass decreases. The extent of the changes is also influenced by the actual deceleration and acceleration rates of the vehicles.

2.3 Chemical Composition

Table 2 presents typical atmospheric concentrations of black carbon (BC), elements and PM mass for different underground railways in different size fractions. It is seen that Fe is a major component in all underground railways

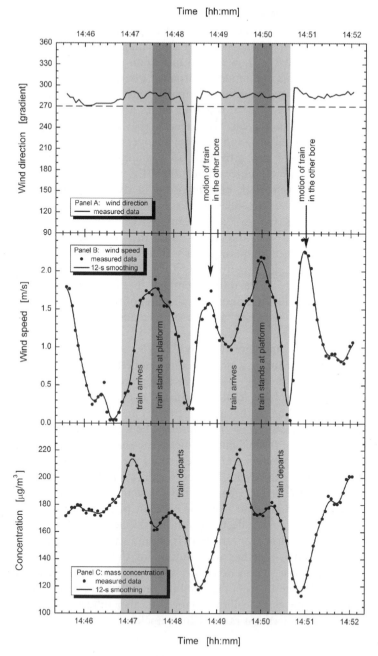

Figure 2 Temporal variation of wind direction (A), horizontal wind speed (B) and PM$_{10}$ mass concentration (C) as measured 4 s averages. The dashed line in A indicates the mean wind direction for moving trains; 12 s smoothed curves are shown in B and C. Time intervals for arriving trains, trains standing at the platform, and departing trains are also displayed.[22] (Reproduced with kind permission of Elsevier.)

Table 2 Typical chemical composition of aerosol particles in $\mu g\,m^{-3}$ for different size fractions in various metropolitan underground railways.

Constituent	Tokyo[9] PM_{10}	Rome[20] PM_{10}	Budapest[22] PM_{10}	Budapest[22] $PM_{2.0}$	Helsinki[18] $PM_{2.5}$	London[4-6,28] —	Mexico[24,28] $PM_{2.5}$	New York[16] $PM_{2.5}$
BC	—	—	3.1	3.1	6.3	—	3.7	—
Mg	0.4	—	0.43	0.13	—	—	—	—
Al	—	—	0.62	0.09	0.27	15.79	—	—
Si	4.9	—	2.5	0.4	0.4	63	2.4	—
S	3.7	—	1.8	0.8	0.7	5.9	8.0	—
Cl	2.3	—	0.41	0.10	0.10	1.3	—	—
K	0.7	—	0.44	0.13	0.19	1.7	0.43	—
Ca	5.2	—	3.0	0.4	0.2	16	0.8	—
Ti	—	—	0.07	0.02	0.03	0.19	0.23	—
Cr	0.60	0.44	0.05	0.01	0.05	0.03	0.10	0.08
Mn	—	0.07	0.46	0.15	0.27	1.7	0.07	0.24
Fe	94	44	49	15	25	87	4.2	26
Ni	0.7	0.07	0.04	0.01	0.03	0.07	0.03	—
Cu	1.0	2.5	0.69	0.19	0.15	0.52	1.6	—
Zn	0.7	0.5	0.17	0.05	0.08	—	—	—
Pb	—	0.09	0.07	0.02	0.01	0.30	0.04	—
PM	—	407	117	33	—	—	31	62

(except for Mexico City where the steel wheels are lined by rubber, see also Section 3). The contributions of Si and Ca are also substantial in the PM_{10} size fraction, while S contributes markedly to the $PM_{2.5}$ mass in addition to the elements listed.

Iron makes up typically about 40% of the PM_{10} mass,[22] while its contribution in the $PM_{2.5}$ size fraction can be as large as 65–70%.[7] The result is comparable with that of 67% obtained for the PM_{10} aerosol generated by railway lines.[33] A somewhat lesser contribution of Fe in subways with respect to that from railway lines implies that additional and significant emission sources of particles are present in metros. Atmospheric concentrations of Fe, Mn, Ni, Cu and Cr in subways are larger (typically by factors of between 50 and 3) than the ambient concentrations above stations. Black carbon concentrations in the $PM_{2.5}$ size fraction for underground railways are typically smaller than, but comparable to, those for average outdoor urban air, while the contribution of BC to the particulate mass is considerably smaller than for urban type aerosol (as expected).[18,22] Atmospheric concentrations of Mn were measured in the Montreal subway[21] and London tube[3] in comparative studies among various transportation modes, in order to investigate if methylcyclopentadienyl manganese tricarbonyl (MMT) added to gasoline or diesel fuels cause higher Mn exposure. Average concentrations of respirable ($PM_{4.3}$ size fraction) Mn in subway stations in Montreal that were located near high surface automobile traffic densities were associated with the outdoor levels, and no correlation was found between the subway traffic and atmospheric Mn in the subway. In London, Mn exposures in the underground microenvironment were approximately two orders of magnitude larger than the exposures of individuals who did not commute *via* the underground.

It is advantageous to compare the enrichment factors (EFs) because they are related to source types and reflect the changes in the chemical composition. Crustal EF of an element X with regard to the reference element R is defined as:

$$EF = \frac{\left(C_X/C_R\right)_{aerosol}}{\left(C_X/C_R\right)_{rock}}, \tag{1}$$

where C_X and C_R represent the concentration of element X and R in aerosol particles (aerosol) and average crustal rock (rock).[32] Of the geogenic elements measured, Si was chosen as the reference elements because it was available for most subways. Crustal EFs of 10^1 imply crustal origin of the element X, while an enrichment significantly larger than this value and than the EF for PM mass indicates non-crustal or anthropogenic origin. It is worth noting that the ratio of the crustal EF for subways to the corresponding crustal EF for the ambient aerosol above the stations (indoor enrichment factor) represents the enrichment in the subway relative to the ambient air. Crustal EFs for Cu, S, Pb, Cl, Zn, Fe, Ni, Cr and Mn are usually the largest ($> 10^2$) in metros. More importantly, Fe, Mn, Ni, Cu and Cr are in general the most enriched elements in subways compared to ambient urban air. Their indoor enrichment factors range from 8 (Cr) to 30 (Fe). Nickel was not enriched in the New York City subway which can most likely be explained by the different types of steel and iron alloys used there (see Section 3). Atmospheric or mass concentrations of these metals are small (except for Fe) but they deserve special attention because of their biological activity.

Little is known about the organic aerosol in underground railways although its contribution to the $PM_{2.5}$ mass could be as large as 50–60% in systems where rubber-lined wheels are in use.[24] Of them, particles of biological origin or microorganisms can be of special interest for epidemiological and antiterrorist considerations, while PAHs are recognised for their carcinogenic health effects. Correlations between the number of microorganisms and the number of passengers or space size are established, and the issue is related to the filtration and air conditioning systems applied.

2.4   Size Distributions

Number size distributions of particles with a diameter below approximately 500 nm in metros are similar to those measured at urban sites above the stations.[18] Less information is available for mass size distributions. Some useful information can be derived from $PM_{2.5}$:PM_{10} or $PM_{2.5}$:$PM_{10-2.5}$ concentration ratios for the particle mass and elements. The $PM_{2.5}$:$PM_{10-2.5}$ ratios for the mass and most constituents in metros are generally smaller than 1.0, implying that the particles and their constituents are mainly formed by mechanical disintegration. The $PM_{2.5}$:$PM_{10-2.5}$ ratios for the mass (of approximately 0.3) observed in several studies indicate that approximately 70% of the particulate mass is formed by coarse particles.[8,22,39] This emphasizes the relative importance of the disintegration emission sources in subways. The ambient aerosol particles found in large cities are generally finer than this. The typical proportion of the $PM_{2.5}$ size fraction in the PM_{10} mass is 60–80% in European cities. In some other studies, however, the $PM_{2.5}$:PM_{10} mass ratios in subways are substantially larger (between 0.6 and 0.8)[12,25,26] than the typical value given above. Several ideas were formulated to explain the larger contribution of the fine particles with regard to the coarse particles. We think that the broadening and shifting of the coarse mode to smaller diameters can contribute significantly to these larger ratios. This explanation is consistent with findings that coarse particles in underground railways are more evenly distributed over the size interval than coarse atmospheric particles.[7] In addition, very strong correlations are usually observed between PM_{10} and $PM_{2.5}$ masses which also imply their common sources.[18] Further and direct investigations are definitely required in this field. In vehicles, the $PM_{2.5}$:PM_{10} mass ratios are generally larger (usually between 0.7 and 0.8) than the values outside the trains.[25] The most likely explanation is that the air conditioning or filtration systems applied on the carriages retain coarse particles more efficiently than fine particles.

2.5   Morphology and Speciation

Images of particles obtained by scanning or transmission electron microscopy methods typically show irregular and angular shapes (see Figure 3 as an example). This is characteristic of particles originating from primary emission sources.[7] The major constituents in most particles are Fe and O. Additional types observed are particles containing large concentrations of Ca, Mg and C; particles containing large concentrations of Si, Ca, Mg, (Cl) and Fe; particles containing large concentrations of Si; and particles containing large concentrations of C. The composition of the groups are consistent with particles of iron oxides, carbonate, silicate, quartz and carbonaceous matter (including plastic debris), respectively. A Fe-rich particle is displayed in Figure 4. The picture reveals that these particles are structured, and they are aggregates of nano-sized crystal grains. The grains are of round shape, have a uniform diameter between approximately 5–15 nm, and they are randomly oriented. The origin and importance of the structure is to be further investigated.

Selected-area electron diffraction pictures show annular pattern which confirm the polycrystalline structure, and imply that the nanocrystals form random conglomerates without specific orientation. The diffraction rings obtained can be utilized to determine the type of the crystal structures and lattice spacings (Miller indices). Two crystal structures were identified in Fe-rich particles: hematite and magnetite. The hematite nanocrystals are more abundant than the magnetite nanocrystals, and in few cases, they occur internally mixed within

| Acc.V | Spot | Magn | Det | WD | Exp | | 5 μm |
| 20.0 kV | 4.0 | 5000x | GSE | 11.1 | 29859 | 118 Pa | metro durva |

Figure 3   Secondary electron image of particles in the PM$_{10-2.0}$ size fraction obtained by scanning electron microscopy.[40] (Reproduced with kind permission of Elsevier.)

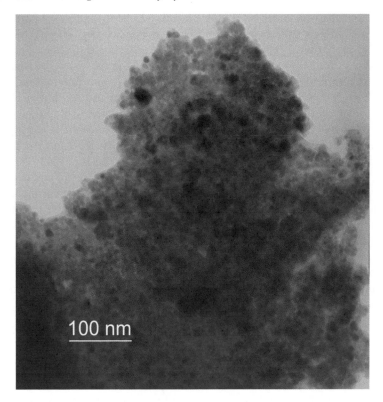

Figure 4   Image of a Fe-rich aerosol particle obtained by transmission electron microscopy showing nano-sized crystal grains within the particle.[40] (Reproduced with kind permission of Elsevier.)

aerosol particles.[40] In some other studies, magnetite dominates.[28] Hematite contains Fe(III), while magnetite contains both Fe(III) and Fe(II). Hematite is the major component of rust. Nevertheless, it is unlikely that the hematite nanocrystals are formed by corrosion processes because iron(III) oxyhydroxides (such as goethite and lepidocrocite) that should also have been formed in the reaction with water vapour are not identified in the $PM_{2.0}$ size fraction, and because corrosion is expected to generate coarse aerosol particles. Instead, it is thought that these aggregated Fe-rich particles are emitted by sparking between the electric conducting rail and collectors.

Information on the chemical forms and speciation, as well as the water solubility (bioavailability), of the other transition metals of interest is of major importance for their health and environmental effects. For instance, Cr(III) is a species essential for life, while Cr(VI) is toxic.[36] A recent study indicates that water solubility of Cr in the PM_{10} size fraction for an underground railway is about 5–6 times smaller than for the city centre.[40] At the same time, the atmospheric concentration of Cr in the metro is larger by a similar factor than for the ambient air. This means that the water soluble amounts for the two environments are similar, and that the increased negative health effects of the aerosol particles in metros with respect to ambient outdoor particles (see Section 4) should be linked to further properties. The water solubility can be also related to source processes. The $PM_{10-2.0}$-fraction Cr is partially soluble, while Cr in the $PM_{2.0}$ size fraction is dissolved completely. The insoluble Cr in the $PM_{10-2.0}$ size fraction likely contains Cr_2O_3 and Cr in its elemental form. Chromium as an alloying element imparts to steels their resistance to corrosion by formation of a thin protective film of Cr_2O_3. These passive films are stable in normal atmospheric and aqueous environments and, therefore, the Cr_2O_3 on the steel particles could remain partially insoluble. However, Cr in the $PM_{2.0}$-fraction particles is partially generated via vaporization by sparking between the electric conducting rail and collectors which can oxidize Cr(0) or Cr(III) emitted from the steel into their higher oxidation states. In the $PM_{10-2.0}$ size fraction, practically all dissolved Cr has an oxidation state of three. This corresponds to ambient conditions where Cr(III) occurs naturally. In the $PM_{2.0}$ size fraction, however, approximately 7% of the dissolved Cr is present as Cr(VI),[40] which can be of concern since Cr(VI) is one of the most harmful metallic components. The differences in the oxidation states and surface properties or morphologies are likely related to the increased harmful health effects of the underground particles.

3   Emission Sources

The crustal EFs for Mg, Si, K and Ti (of roughly 10^0) for the aerosol in metros strongly suggest that the emission sources for these elements are most likely disintegration, dispersal and resuspension of the crustal rock, concrete and ballast due to normal operation and construction work in the tunnels and stations. Wind erosion of construction materials, and of the materials covering the surfaces may also play a role. For the other elements in Table 2 and PM mass, the crustal EFs are much greater than 10^0, indicating that their main emission sources are of non-crustal origin. The largest indoor EFs for Fe, Mn, (Ni,) Cu and Cr, their association with the $PM_{10-2.0}$ size fraction, and their strong correlation imply that these metals are primarily emitted into the air within subways from common non-crustal sources by mechanical disintegration.[7,8,22] Elemental ratios (signatures) help to trace potential sources. The Fe:Mn, Fe:Cr and Fe:Ni concentration ratios for the aerosol particles in metros (typical of approximately 110, 1000 and 1100, respectively) are significantly different from the ratios (of 500, 53, and 667, respectively) for the average crustal rock composition,[32] and are similar to the typical ratios obtained for the alloy steel used for rails and electric collectors.[3,7,16,18,22] Taken together, emissions by friction and rubbing of the electric rails and plough collectors, and by wear of steel wheels on steel rails are the major source of PM_{10} mass, Fe, Mn and Cr. This is demonstrated in Figure 5 by a scatter plot of the 30 min mean PM mass and the number of trains passing the platform over 30 min intervals.[22] Metros with rubber-lined steel wheels have considerably lower Fe concentration levels than the metros that utilise ordinary steel wheels. Friction of these

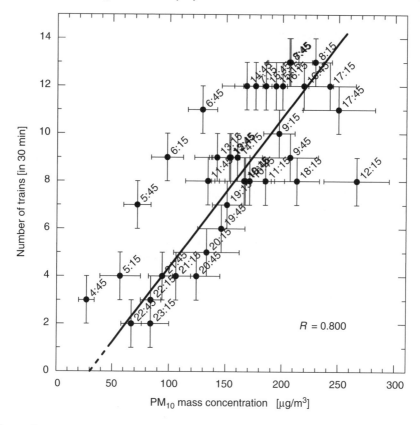

Figure 5 Scatter plot of the PM_{10} mass concentration versus the number of trains that passed the metro station in the direction examined for 30 min periods. The correlation coefficient (*R*) is also indicated.[22] (Reproduced with kind permission of Elsevier.)

wheels is an important source of coarse organic aerosol and S, and the organic aerosols can become a major component in such networks.[24]

For the $PM_{2.5}$ size fraction, crustal and indoor EFs are greater than or similar to those for the PM_{10} size fraction in many cases. This is consistent with the interpretation of a broader size distribution for coarse particles as outlined in Section 2.4. Indoor EFs for BC, S and K in the $PM_{2.5}$ size fraction suggest that these aerosol constituents could be transported with outdoor air through ventilation systems, leading to an observable depletion of BC. Moreover, BC correlates well with that at the urban sites nearby, while its correlation with the $PM_{2.5}$ mass is generally poor, indicating that the main sources of BC for subways are the traffic exhaust fumes on the surface. Some BC can also be emitted from the carbon brushes of the driving electric motors or by diesel-driven work cars used for service activities overnights. Other potential sources of metals include contact points for electric switches, vaporisation due to sparking involving the third rail, and composite brake pads. Passengers travelling in subways and evaporation of some organic compounds from detergents, wood preservatives or paints followed by their aging and condensation can in principle contribute to organic aerosol as well. In the outdoor air, pollutant trace gases (mainly organics) undergo photochemical reactions usually initiated by OH radical (effective during daylight periods) or by NO_3 radical (effective overnight) which often transform them into the aerosol phase. Formation of these radicals and several relevant chemical reactions in the air are governed by sunlight. These processes are thought to be of much less (mostly negligible) importance in the underground environments than in the outdoor air. This also contributes to the differences in chemical composition. Ultrafine particles in underground systems usually originate from outside and not from subways, and are related to the introduction of outdoor air that contains pollutants from traffic and other high-temperature sources on nearby streets.

Further research is needed to clarify and establish the relative contribution of various sources to the PM and its chemical composition.

4   Consequences for Human Health

Exposures to high levels of aerosol articles are associated with excess morbidity and mortality for outdoor conditions. Relatively little is known about potentially related health effects in the metro systems even though most studies suggest that the mass concentration levels are higher than the outdoor levels. Very few toxicological studies have been conducted, and no epidemiological studies have been published. The health implications for passengers and workers in metros can be approached differently. As far as the public is concerned, the PM_{10} and $PM_{2.5}$ concentrations for many underground railways listed in Table 1 are several-fold greater than the ambient 24 h limits. It is recognized that the average commuter spends a small fraction of 1 day in underground spaces in walking to, waiting for and riding trains on a daily basis, and, therefore, the high aerosol concentrations observed have mainly short-term health effects. Higher health risks for sensitive groups, such as children, the elderly and individuals with pre-existing health conditions exacerbated by air pollution (many respiratory and cardiovascular diseases) may be significant, even for the short times spent in underground railways. The time spent on platforms (or in stations) may be a better predictor of personal exposure than total time spent underground (see Section 2.1). Recent studies suggest that the short-term exposure is associated with acute adverse health effects, even at low levels (0–100 μg m^{-3}) of exposure.[35] Moreover, the impact is often repeated almost every day for most commuters, which may cause cumulative, quasi-long-term or chronic health effects over time. It was estimated that PM exposure increased by several percent up to several tens of percent for normal travel routines in underground railways, while the increase for the transition metals listed can be very large.[3,4,19,24] It should also be considered that the chemical composition and properties of indoor aerosol particles differ from those for outdoor air. The particles are larger and heavier, are mainly composed of Fe, and contain less BC, and, therefore, their impact on morbidity and mortality is expected to be different. Moreover, a large mass fraction of these particles is deposited fortunately in the upper airways of the human respiratory system. The fraction of the inhaled aerosol initially deposited is modified by several clearance mechanisms operating with different magnitudes in the various regions of the respiratory tract. For the larger particles in subways, a larger portion of the deposited mass appears eventually in the extrathoracic region. From this region, the deposited particles generally enter into the gastro-intestinal tract, which in general, represents a smaller health risk than the lungs because of less efficient uptake. It would be inappropriate or over-simplistic to estimate the health risk due to increased particulate mass only.

Exposure to the transition metals that are highly enriched relative to both average crustal rock and average outdoor air may represent a major proportion of the daily personal exposure.[3,16–18] Underground railways seem to be the dominant microenvironment responsible for the exposure to Fe, Cr, Mn and possibly Ni for persons who are not subject to occupational exposure. The health effects of these metals have been identified from occupational intoxication or epidemiological studies at high concentration levels that exceed ambient health limits. The harmful health impacts of the excess exposure to Fe (and other transition metals) is due—in large part—to free iron ions on the surface, available to take part in Fenton chemistry and to generate free radicals in the body, resulting in oxidative stress, inflammatory reactions, neurode-generative diseases, and multiple sclerosis.[34,38] Chromium and Ni are known to be carcinogenic, especially in the respiratory organs, and Cr can lead to gene mutations. Manganese is classified as neurotoxic and causes respiratory dys-functions such as pneumonia.[37] Chromium, Mn and Ni were identified as hazardous air pollutants (air toxics), and their emissions are controlled by regulatory measures in most countries. The health effects of these metals at intermediate concentrations (comparable to the ambient outdoor limit, but lower than the workplace limit) are unknown, similarly to the cumulative impact of their concurrent exposures.

It was reported recently that particles from metros are eight times more genotoxic and four times more likely to cause oxidative stress than particles from busy urban streets as well as other particle types.[38] Substances that are genotoxic may bind directly to DNA or act indirectly leading to DNA damage by affecting enzymes involved in DNA replication, thereby causing mutations which may or may not lead to cancer or birth defects (inheritable damage). The genotoxicity of subway particles was investigated by comparing the ability of these particles and particles from a street, pure tyre-road wear particles, and particles from wood and diesel combustion to form intracellular reactive oxygen species. The genotoxicity and ability to cause oxidative stress was also compared to particles consisting of iron oxide since this is a main component in subway particles. It was concluded that the subway particles and also street particles and particles from wood and diesel combustion can damage the mitochondria, and, therefore, this is not the only explanation for the high genotoxicity of subway particles. Subway particles also form intracellular reactive oxygen species. This effect may be part of the explanation as to why subway particles show high genotoxicity when compared to that of other particles. Genotoxicity can not, however, be explained by iron oxide (magnetite), by water-soluble metals, or by intracellular mobilized iron. The genotoxicity is most likely caused by highly reactive surfaces (by the location and coordination of the metals or perfection of the crystals on the particle surfaces) giving rise to oxidative stress.

Train drivers and other workers who spend several hours a day within underground metros are subject to greater exposure to steel dust than the commuting public, and thus greater health risks. Health regulations for workplace environments apply to such workers. The concentrations in underground railways are clearly lower (by several orders of magnitude) than the corresponding workplace limits or guidelines. For comparison, 8 h average maximum permissible concentration for inhalable (PM_{10} size fraction) inert or nuisance dust is 10 mg m^{-3} in many European countries.

5 Conclusions

One of the most severe problems in large cities is the surface traffic which is linked to general overcrowding, traffic congestion and consequently to concentrated air pollution, noise contamination and other environmental, health and climatic impacts. Complex actions and integrating assessments that can manage the issue of the urban traffic can solve or decrease the extent of these troubles as well. The use of high-capacity and fast modes of transport is indispensable in urban areas for reasons of social, economic and environmental sustainability. Despite the high investment costs, the development potential of underground railway systems is still large since some 560 cities with populations of over one million are expected by 2015. The existing metro lines are also to modernise which is not only a technological challenge but also requires a complete rethinking of service philosophy and health issues. In parallel, it is definitely required that the number of studies on subway-related air pollution and exposure, on comparison of alternative modes of urban public transportation, as well as on source-specific epidemiological investigations is increased. There is still much to learn about the air pollution in underground railway systems and about the pollutants that metro systems emit to their environment. It is clear and mandatory that the air quality of underground railways is improved. At the same time, we emphasise again that there are no known health effects at the concentrations levels measured in metros, and therefore, the present summary in no way suggests that travellers should avoid riding the subways. The existing conclusive knowledge and some perspective results in this field can be used on the one hand for avoiding exacerbation of anxieties for the public, and on the other hand for helping in the selection of advanced construction materials and in informing improved design, creating new lines or in turning the present systems into cleaner and more attractive public transport facilities.

Acknowledgements

This work was partially supported by the Hungarian Scientific Research Fund (contract K061193).

References

1. Public Transport International, *Metro: A Bright Future Ahead*, International Association of Public Transport, Brussels, 2007.
2. H. Fromme, A. Oddoy, M. Piloty, M. Krause and T. Lahrz, *Sci. Total Environ.*, 1998, **217**, 165.
3. G. D. Pfeifer, R. M. Harrison and D. R. Lynam, *Sci. Total Environ.*, 1999, **235**, 253.
4. H. Adams, M. Nieuwenhuijsen, R. N. Colvile, M. McMullen and P. Khandelwal, *Sci. Total Environ.*, 2001, **279**, 29.
5. H. S. Adams, M. J. Nieuwenhuijsen and R. N. Colvile, *Atmos. Environ.*, 2001, **35**, 4557.
6. H. S. Adams, M. J. Nieuwenhuijsen, R. N. Colvile, M. J. Older and M. Kendall, *Atmos. Environ.*, 2002, **36**, 5335.
7. B. Sitzmann, M. Kendall, J. Watt and I. Williams, *Sci. Total Environ.*, 1999, **241**, 63.
8. A. Seaton, J. Cherrie, M. Dennekamp, K. Donaldson, J. F. Hurley and C. L. Tran, *J. Occup. Environ. Med.*, 2005, **62**, 355.
9. K. Furuya, Y. Kudo, K. Okinaga, M. Yamukki, S. Takahashi, Y. Araki and Y. Hisamatsu, *J. Trace Microprobe Techn.*, 2001, **19**, 469.
10. A. H. A. Awad, *J. Occu. Health*, 2002, **44**, 112.
11. L. Y. Chan, W. L. Lau, S. C. Lee and C. Y. Chan, *Atmos. Environ.*, 2002, **36**, 3363.
12. L. Y. Chan, W. L. Lau, S. C. Zou, Z. X. Cao and S. C. Lai, *Atmos. Environ.*, 2002, **36**, 5831.
13. A. Birenzvige, J. Eversole, M. Seaver, S. Francesconi, E. Valdes and H. Kulaga, *Aerosol Sci. Technol.*, 2003, **37**, 210.
14. Ch. Johansson and P-Å. Johansson, *Atmos. Environ.*, 2003, **37**, 3.
15. G. Axelsson, *Dust Measurements in the Underground of Stockholm,* Report of Arbetsmiljögruppen, Stockholm, 1997.
16. S. N. Chillrud, D. Epstein, J. M. Ross, S. N. Sax, D. Pederson, J. D. Spengler and P. Kinney, *Environ. Sci. Technol.*, 2004, **38**, 732.
17. S. N. Chillrud, D. Grass, J. M. Ross, D. Coulibaly, V. Slavkovich, D. Epstein, S. N. Sax, D. Pederson, D. Johnson, J. D. Spengler, P. L. Kinney, H. J. Simpson and P. Brandt-Rauf, *J. Urban Health*, 2005, **82**, 33.
18. P. Aarnio, T. Yli-Tuomi, A. Kousa, T. Mäkela, A. Hirsikko, K. Hämeri, M. Räisänen, R. Hillamo, T. Koskentalo and M. Jantunen, *Atmos. Environ.*, 2005, **39**, 5059.
19. M. Braniš, *Atmos. Environ.*, 2006, **40**, 348.

20. G. Ripanucci, M. Grana, L. Vicentini, A. Magrini and A. Bergamaschi, *J. Occup. Environ. Hygiene*, 2006, **3**, 16.
21. N. Boudia, R. Halley, G. Kennedy, J. Lambert, L. Gareau and J. Zayed, *Sci. Total. Environ.*, 2006, **366**, 143.
22. I. Salma, T. Weidinger and W. Maenhaut, *Atmos. Environ.*, 2007, **41**, 8391.
23. T.-T. Li, Y.-H. Bai, Z.-R. Liu and J.-L. Li, *Transportation Research, Part D*, 2007, **12**, 64.
24. J. E. Gomez-Perales, R. N. Colvile, A. A. Fernandez-Bremauntz, V. Gutierrez-Avedoy, V. H. Paramo-Figuero, S. Blanco-Jimenez, E. Bueno-Lopez, R. Bernabe-Cabanillas, F. Mandujano, M. Hidalgo-Navarro and M. J. Nieuwenhuijsen, *Atmos. Environ.*, 2007, **41**, 890.
25. Y.-H. Cheng, Y.-L. Lin and Ch.-Ch. Liu, *Atmos. Environ.*, 2008, **42**, 7242.
26. K. Y. Kim, Y. S. Kim, Y. M. Roh, Ch. M. Lee and Ch. N. Kim, *J. Hazard. Mater.*, 2008, **154**, 440.
27. D.-U. Park and K.-Ch. Ha, *Atmos. Environ.*, 2008, **34**, 629.
28. M. J. Nieuwenhuijsen, J. E. Gomez-Perales and R. N. Colvile, *Atmos. Environ.*, 2007, **41**, 7995.
29. Ch.-Ch. Chan, J. D. Spengler, H. Özkaynak and M. Lefkopoulous, *J. Air Waste Manage. Assoc.*, 1991, **41**, 1594.
30. Council Directive 1999/30/EC, *Official Journal of the European Communities*, L163, 41, 1999.
31. *US Environmental Protection Agency National Ambient Air Quality Standards*, 40 CFR, parts 50–96, 1997.
32. B. Mason and C. B. Moore, *Principles of Geochemistry,* Wiley, New York, 1982.
33. R. Lorenzo, R. Kaegi, R. Gehrig and B. Grobéty, *Atmos. Environ.*, 2006, **40**, 7831.
34. K. Donaldson, D. M. Brown, C. Mitchell, M. Dinerva, P. H. Beswick, P. Gilmour and W. McNee, *Environ. Health Perspect.*, 1997, **105**, 1285.
35. Health Effects Institute Review Committee, *Understanding the Health Effects of Components of the Particulate Matter: Progress and Next Steps*, Health Effect Institute, Boston, 2002.
36. World Health Organization, *Air Quality Guidelines for Europe, European series No. 91*, WHO Regional Office for Europe, Copenhagen, 2000.
37. Agency for Toxic Substances and Disease Registry, *Toxicological Profiles*, US Public Health Service, Atlanta, 2000.
38. H. L. Karlsson, Å. Holgersson and L. Möller, *Chem. Res. Toxicol.*, 2008, **21**, 726.
39. J.-C. Raut, P. Chazette and A. Fortain, *Atmos. Environ.*, 2009, **43**, 860.
40. I. Salma, M. Pósfai, K. Kovács, E. Kuzmann, Z. Homonnay and J. Posta, *Atmos. Environ.*, 2009, accepted.

Human Exposure: Indoor and Outdoor

SOTIRIS VARDOULAKIS

ABSTRACT

Human exposure to air pollution is highly variable, reflecting the strong spatial and temporal variability of air pollutant concentrations in urban environments. A wide variety of outdoor and indoor sources contribute to acute and chronic exposures to respirable particles, carbon monoxide, nitrogen oxides, sulfur dioxide, ozone, volatile organic compounds and polycyclic aromatic hydrocarbons. Ambient air quality measurements and dispersion model simulations are commonly used to estimate population exposure to air pollutants in cities. In addition, time-activity microenvironmental and other dynamic time-space models can be used to characterise the personal exposure of individuals or population subgroups. People in developed countries typically spend 90% of their time indoors, and impact of indoor sources, such as tobacco smoking, gas cooking, construction and furnishing materials, and household chemicals (e.g. paints, adhesives, cleaning products, etc.), on personal exposure can become dominant. Recent exposure surveys have shown that personal exposure is typically higher than both indoor and outdoor concentrations of traffic-related pollutants such as benzene. In most cases, this is due to peak personal exposures occurring within transient (e.g. commuting) and other highly polluted micro-environments (e.g. petrol stations, garages, etc.). In developing countries, domestic fuel combustion (e.g. biomass burning for cooking and heating) has been identified as a major factor contributing to elevated exposure of the population to respirable particles, carbon monoxide, polycyclic aromatic hydrocarbons, nitrogen oxides and sulfur dioxide.

Issues in Environmental Science and Technology, 28
Air Quality in Urban Environments
Edited by R.E. Hester and R.M. Harrison
© Royal Society of Chemistry 2009
Published by the Royal Society of Chemistry, www.rsc.org

1 Introduction

Exposure can be defined as the interaction between air pollution and receptors, either individuals or population groups, distributed in space and time. It is important to make a distinction between concentrations, exposure and dose. The concentration of an air pollutant is a characteristic of the atmospheric environment at a certain location and time, while exposure is a characteristic of both the human activity and the environment.[1] For example, high concentrations of an air pollutant may occur downwind of an industrial site where there are no residential properties. In this case, population exposure levels are likely to be very low despite the high ambient concentrations of the pollutant. Human exposure becomes a dose when a contaminant moves across an absorption barrier (*e.g.* respiratory tract lining). To assess health impacts, the information about exposure (or dose) has to be combined with exposure–response (or dose–response) relationships usually derived from epidemiological studies, as indicated in Figure 1.

Exposure to air pollution may occur from multiple indoor (*e.g.* domestic cooking and heating, solvents from consumer products, environmental tobacco smoke, *etc.*) and outdoor sources (*e.g.* road transport, industrial sources, biogenic emissions, *etc.*) over repeated short events or long continuous time periods. People living in developed countries typically spend 90% of their time indoors, with vulnerable individuals (elderly, young children, people with compromised health) spending an even larger proportion of their time at home. Therefore, indoor air pollution may contribute more substantially to personal

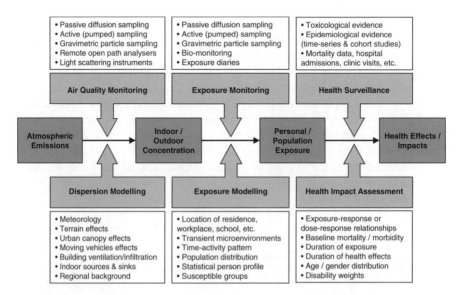

Figure 1   Schematic illustration of the chain of events from air pollutant emissions to health effects, and associated assessment methodologies.

exposure than outdoor air pollution, depending on individual lifestyles and housing conditions.

Some contaminants, such as volatile organic compounds (VOC) and metals, may be present in more than one environmental media (air, groundwater, soil). The sum of exposures to the same contaminant from all sources and pathways (inhalation, dermal contact and ingestion) is usually referred to as *aggregate exposure*, while the sum of exposures over a relatively long period of time is called *cumulative exposure*. The term cumulative exposure has also been used in scientific literature to express exposure to multiple contaminants causing a common health effect. Cumulative exposure assessment (*i.e.* over a long period of time) is particularly important when characterising risks associated with long-latency health outcomes, such as lung cancer and asthma.[2] In the following sections, the main sources and characteristics of outdoor and indoor exposure to air pollution, assessment methodologies and relevant research studies are discussed. Occupational exposure to air pollutants occurring in the workplace is not included in this chapter.

2 Characteristics of Outdoor and Indoor Exposure

2.1 Outdoor Exposure

Air pollution levels in medium and large-size cities are highly variable. This is due to the complexity of the urban canopy and the large amount of emission sources that can be very different in nature and spatial distribution. Road traffic is the main source of air pollution in urban areas, while domestic heating, industry, power generation, fugitive and re-suspended dust emissions, and long-range atmospheric transport also contribute to the ambient pollution levels. From a human exposure point of view, the key outdoor air pollutants are nitrogen dioxide (NO_2), fine particles (PM_{10} and smaller size fractions), carbon monoxide (CO), sulfur dioxide (SO_2), ozone (O_3), lead, benzene, 1,3-butadiene, and certain polycyclic aromatic hydrocarbons (PAH). In practice, NO_2 and PM_{10} are causing most of the exceedences of air quality standards in European cities, with ozone and benzene also causing some concern, mainly in southern Europe. Several markers of exposure to air pollution have been used in the past to assess health risks. NO_2 has been commonly used as an indicator of exposure to traffic-related air pollution due to its relatively well established health effects and easy access to monitoring data and emission factors.[3]

The presence of buildings and other features of the urban topography, such as parks, may enhance or reduce human exposure to air pollution. For example, pollutant concentrations are generally higher inside *street canyons*, *i.e.* busy streets surrounded by relatively high buildings on both sides,[4] compared to unconfined outdoor spaces. Another example is urban vegetation, which can act as an effective sink of airborne particles reducing exposure to PM_{10}. Other localised sources, such as petrol stations, industrial plants, construction sites, *etc.*, may also create air pollution *hotspots*. Several experimental studies have highlighted the strong spatial variability of air pollution in urban areas (*intra-urban variability*), mainly near busy streets.[5–9] It should be noted that the chemical composition of particulate matter (PM) can also vary substantially with distance from the road, as kerbside PM is expected to include a higher proportion of traffic-related species, such as elemental carbon and PAH.[10]

Apart from horizontal variability, vertical pollution gradients may have a significant effect on population exposure in built-up areas.[11] Vertical concentration profiles of non-reactive pollutants in a street canyon generally satisfy a law of exponential reduction with height, although more complex patterns, depending on the side of the street and distance from buildings, have been observed. For secondary reactive pollutants like ozone, this pattern may be reversed with higher concentrations occurring near the top of the canyon.

In addition to spatial variability, temporal and seasonal trends may influence population exposure levels. Continuous air quality monitoring of traffic-related pollutants, such as CO and NO_2, has shown that pollutant concentrations follow closely the daily, weekly and seasonal road traffic pattern in urban areas. As expected, peak concentrations occur during rush hours, while pollution levels are generally lower during night time, weekends and school holiday periods. Finally, the local meteorology and weather cycle affect ambient air pollution and consequently population exposure levels. Winter smog episodes (*i.e.* elevated NO_2, SO_2 and PM_{10} concentrations) induced by poor atmospheric dispersion conditions, heavy road traffic and fuel combustion for domestic heating, have been associated with increased mortality and morbidity in European and North American cities during winter. On the other hand, summer smog episodes (*i.e.* elevated NO_x and O_3 concentrations) may exacerbate respiratory symptoms in urban and sub-urban areas on hot and sunny summer days.

2.2   Indoor Exposure

Previous studies have shown that personal exposure may correlate more strongly with indoor pollution levels than with outdoor concentrations, especially in homes with indoor combustion sources.[12] Although indoor air quality levels are also highly variable, depending on emission sources and ventilation rates, there is strong evidence that certain pollutant concentrations (*e.g.* benzene) in houses may exceed ambient levels in European cities.[13,14] In developing countries, the domestic combustion of fuel for heating and cooking makes a substantial contribution to personal exposure (mainly of the female population) to respirable particles, CO, PAH, NO_2 and SO_2.[15]

Indoor emission sources associated with personal exposure are very diverse. They can be broadly classified into the following categories: (a) emissions from building and furnishing materials, including asbestos, formaldehyde and persistent organic pollutants (PCB, PBDE, *etc.*); (b) emissions from household consumer products, such as solvents (VOC), pesticides and insecticides; (c) fuel and tobacco combustion emissions, including volatile and semi-volatile organic compounds, CO, $PM_{2.5}$, NO_2 and SO_2; (d) O_3 from photocopiers and other electronic appliances; (e) biological pollutants such as mould, bacteria and viruses and dust mites. In addition to the above sources, pollutants can infiltrate into the indoor environment from the soil under the building (radon gas, methane and other landfill gases) and from outdoors (traffic-related pollutants, ozone and pollen). Some of these pollutants (gases and vapours) can be absorbed and possibly re-emitted by building and furnishing materials, while particles settle on indoor surfaces from where they can easily be re-suspended.

In indoor residential environments, emphasis has traditionally been put on short-term exposure to CO, which can cause intoxication and acute morbidity when accumulated in confined spaces. Although the effects on health of low level exposure to CO are still unclear, there is increasing evidence that long-term exposure to low CO concentrations may have a significant impact by causing neurological damage.[16]

2.3 Time–Activity Patterns

Human beings are "dynamic" receptors, spending different proportions of their time in a variety of *micro-environments*, *i.e.* spaces with uniform pollutant concentrations (*e.g.* kitchen, bedroom, office, car, park, street, restaurant, supermarket, *etc.*). The exposure of an individual to air pollution depends on the frequency and amount of time spent in each one of these micro-environments, the pollutant concentration in this micro-environment, and the uptake (*i.e.* breathing) rate. Strictly speaking, the distribution of time to daily activities (*time–activity pattern*) is different for each individual, but in practice certain common patterns can be identified, for example for working or retired adults, pre-school children or children attending school, commuters, *etc.* Human exposure also depends on a range of factors affecting breathing rates, such as age, gender, weight, physical condition and activity level. For example, the typical breathing rate ($1\,day^{-1}$) for adult males is usually higher than for adult females. Typical breathing rates are lower for children compared to adults, although contaminant uptake per body weight is generally larger for children.

2.4 Indoor–Outdoor Relationships

A common assumption in exposure studies is that air pollutant concentrations are uniform within residential environments that do not comprise particularly strong indoor sources such as tobacco smoking, wood burning, *etc.* This is due to the generally more effective air mixing between the rooms of a house, compared to the air exchange between indoors and outdoors. It can also be assumed that indoor concentrations of traffic-related pollutants in non-smoking households correlate strongly with outdoor concentrations dominated by road traffic emissions.[17] Therefore, population exposure to these pollutants can be roughly estimated from their ambient levels and indoor : outdoor (I : O) concentration ratios observed in similar urban environments. Typical pollutant concentrations and I : O ratios are reported in Table 1.

Depending on residence location, outdoor concentrations of traffic-related pollutants (*e.g.* NO_2) may be higher than indoor levels. On the other hand, indoor concentrations of pollutants that are associated with household activities and products (VOC) may be substantially higher indoors than outdoors. A range of common indoor activities have been associated with increased exposure to specific VOC, such the use of deodorisers (*p*-dichlorobenzene), washing clothes and dishes (chloroform), tobacco smoking (benzene and styrene) and painting (*n*-decane).[18]

3 Exposure Assessment

Assessing exposure is an important aspect of any risk analysis and health impact assessment study. There are two broad types of assessment methodologies, direct and indirect. The direct approach may involve individuals carrying sampling devices that monitor pollutant concentration within their breathing zone for certain period of time. Indirect assessment methods often use a surrogate, such as proximity to the source or ambient pollutant concentrations, to estimate exposure. The most commonly used exposure monitoring and modelling techniques are discussed in the following sections.

3.1 Exposure Monitoring

3.1.1 Monitoring Networks
Air quality monitoring networks providing continuous measurements of key pollutants (*e.g.* NO_x, PM_{10}, $PM_{2.5}$, SO_2, CO) are implemented in many medium and large-size cities in Europe, North America and elsewhere. Some of the most commonly used automatic analysers in monitoring networks are summarised in Table 2. Specific EU guidelines require that sampling points be located within the human breathing zone (1.5–4 m above the ground), at least 25 m away from the edge of any major junctions and at least 4 m from the centre of the nearest traffic lane.

The spatial coverage, design and density of automatic urban monitoring networks vary widely. Fixed automatic monitoring stations are expensive to set up and maintain, and the number of such stations is therefore limited within a city. Passive sampling has been used as an alternative low-cost air quality monitoring technique. Passive (or diffusion) samplers are generally easy to deploy and do not require power supply, but they tend to be less accurate than automatic gas analysers. They are based on the principle of molecular diffusion, rather than pumped sampling, for drawing molecules of the contaminant onto the collection medium (*e.g.* chemical adsorbent material) of the sampling device. Nowadays, passive sampling networks are implemented in many urban areas, mainly focusing on NO_2 and benzene. In the UK, there are over 1500 permanent monitoring sites of which only 128 sites operate automatically,

Table 1 Indoor : outdoor concentration ratios reported in literature. [49,61-63]

Pollutant	Median concentrations		Ratio I:O	Sampling technique	Reference	Country	Study
	Indoor	Outdoor					
Formaldehyde ($\mu g/m^3$)	20.10	6.42	3.1	Passive sampler (PAKS)	Liu et al. 2006	USA (homes)	RIOPA
$PM_{2.5}$ ($\mu g/m^3$)	23.2	23.4	1.0	Harvard impactors (Teflon filters)	Janssen et al. 2001	Netherlands (schools)	
NO_2 ($\mu g/m^3$)	17.5	38.5	0.5	Passive diffusion tubes (Palmes)	Janssen et al. 2001	Netherlands (schools)	
Benzene ($\mu g/m^3$)	2.9	2.1	1.4	Pumped charcoal tubes (SKC)	Janssen et al. 2001	Netherlands (schools)	
Benzene ($\mu g/m^3$)	1.48–2.17	1.13–1.62	1.2–1.4	Passive sampling badges (3M OVM 3500)	Schneider et al. 2001	Germany (homes)	INGA
Toluene ($\mu g/m^3$)	20.46–37.29	4.46–4.98	5.0–7.6	Passive sampling badges (3M OVM 3500)	Schneider et al. 2001	Germany (homes)	INGA
m,p-Xylene ($\mu g/m^3$)	2.92–4.17	1.20–1.76	2.1–1.8	Passive sampling badges (3M OVM 3500)	Schneider et al. 2001	Germany (homes)	INGA
Benzo[a]pyrene* (ng/m^3)	0.09	0.10 ng/m³	0.9	GRAVIKON PM 4 mobile dust sampler	Fromme et al. 2004	Germany (homes)	
Benzo[a]pyrene** (ng/m^3)	0.27	0.10 ng/m³	2.7	GRAVIKON PM 4 mobile dust sampler	Fromme et al. 2004	Germany (homes)	

* Non-smoker
** Smoker

Table 2 Devices commonly used in exposure monitoring.

Pollutant	Automatic (fast-response) sampling	Integrating (pre-concentration) sampling
NO_2	Chemiluminescence gas analysers	Palmes diffusion tube sampler (TEA absorbent)
CO	Infrared gas analysers	Not commonly used
SO_2	Fluorescence gas analysers	Diffusion tube (potassium hydroxide adsorbent), bubbler and filter pack methods
O_3	Ultraviolet gas analysers	Ogawa badge sampler (sodium nitrite adsorbent)
Hydrocarbons	Automatic gas chromatography (e.g. VOCAIR)	Perkin-Elmer BTX diffusion tube (Tenax TA absorbent), RADIELLO BTX cartridge sampler (Carbograph adsorbent), 3M 3500 badges-type OVM sampler (charcoal adsorbent)
PAH	Not commonly used	Polyethylene-based passive sampler devices
$PM_{10}/PM_{2.5}$	Tapered Element Oscillating Microbalance (TEOM), and Beta attenuation monitors	Pumped gravimetric sampling (e.g. Partisol monitor)

while most of the rest belong to the NO_2 passive diffusion tube monitoring network.[19] Sampling of SO_2, polycyclic aromatic hydrocarbons (PAH) and heavy metals using manual procedures is also carried out at a limited number of sites across the country.

The main advantage of automatic air quality monitoring networks is that they provide reliable measurements of key pollutants at a high temporal resolution (*i.e.* hourly). This type of dataset can be used in time-series epidemiological studies aiming to establish associations between air pollution and short-term health effects. On the other hand, passive sampling can be used to characterise spatial gradients of air pollution, helping to establish discrete exposure zones at a relatively low temporal resolution (typically one to several weeks).

It should be remembered that ambient air quality measurements are often site-dependent and not representative of population exposure. This is particularly the case for measurements obtained from monitoring stations located near busy streets and junctions in built-up areas.[20,21] Experimental studies carried out in Paris, France[9] and Bologna, Italy[22] have demonstrated that urban air pollution data from fixed air quality monitoring stations cannot automatically be taken as indicators of exposure.

3.1.2 Personal Monitoring

A variety of active (*i.e.* pumped instruments) and passive devices (*e.g.* diffusion tubes) have been used to monitor personal exposure to air pollution. These devices can be either integrating or fast response instruments. Integrating (also called *pre-concentration*) monitoring techniques collect gaseous pollutants or particles on an appropriate adsorbent bed or filter, respectively, which can be analysed or weighted later in a laboratory. Fast response monitoring may rely on optical or electrochemical techniques to record pollutant concentrations at very high temporal resolution (*e.g.* one second). Integrating monitoring has been commonly used in personal exposure studies, while fast response instruments are now becoming more popular.

Ideally, air samples should be collected within the breathing zone of individuals to reflect true personal exposure *via* inhalation. However, in practice, personal samplers connected to pumps may have to be carried by subject volunteers in briefcases or rucksacks during the sampling campaign. Passive diffusion sampling is a practical low-cost technique for assessing personal exposure to air pollutants, such as NO_2, benzene and other hydrocarbons. Passive samplers have been extensively used for this purpose in occupational and residential environments.[3,23] Several types of passive diffusion samplers and adsorbents can be used depending on the target pollutant and period of exposure. Badge-type samplers have typically higher uptake rates which make them suitable for measuring relatively short-term exposure (*e.g.* daily), while tube-type samplers have lower uptake rates and are therefore more suitable for measuring relatively long-term exposures (*e.g.* weekly to monthly). Personal monitoring can be used in cross-sectional or case-control epidemiological studies that seek to establish a relationship between exposure and prevalence or incidence of disease.

Pumped VOC sorbent tubes, gravimetric instruments collecting particles on filters, passive electrochemical CO sensors, automatic particle counters, and other devices have also been used to monitor personal exposure in a variety of environments.[24,25] Active sampling systems have the capability of mimicking human inhalation rates (20 m^3 day^{-1}). Well calibrated pumped sorbent tubes typically produce measurements of higher accuracy than equivalent passive samplers. On the other hand, active sampling systems are usually heavier and bulkier (including sampling pump), complex to set up, calibrate and maintain, and substantially more expensive.

An alternative method for assessing human exposure in epidemiological studies is by means of questionnaires and/or interviews. This indirect monitoring technique has been used to collect information on presence, duration, frequency and pattern of exposure to air pollution (*e.g.* environmental tobacco smoke). Although less accurate than personal sampling, questionnaire surveys are low cost and can reach a large number of subjects, which increases the statistical power of epidemiological analysis.[26]

3.1.3 Biomonitoring

Biomonitoring is a direct method for estimating human exposure to air pollutants which accumulate in certain parts of the body, or generate a range of biochemical and physiological responses. That involves the use of a *biomarker* which may be the inhaled contaminant (*direct biomarker*) or a metabolite of this substance (*indirect biomarker*).

Biomonitoring is based on the identification, sampling and analysis of appropriate biomarkers that reflect current body burden, which can be associated with recent and/or past exposure. Samples may include blood, urine, saliva, hair, bone, teeth, nail, *etc.*, of the exposed individual. Collection of biomarkers can be either invasive (*e.g.* blood sampling) or non-invasive (*e.g.* urine sampling), which is preferable for long-term or repeated monitoring. The main advantage of biomonitoring is that only the contaminants that cross an absorption barrier and enter the human body are measured. Furthermore, it helps estimate aggregate exposure, as all exposure pathways are included (inhalation, dermal contact and ingestion). On the other hand, the main limitation of biomarkers is their lack of specificity, which means that they cannot be used to identify a specific air pollution source in areas where multiple sources are present. Laboratory analysis of biomarkers is usually expensive and time-consuming.

Biomarkers have been used to estimate human exposure to various air pollutants, such as lead, CO, benzene and other VOCs. In the case of lead, for example, short-term exposure can be estimated through analysis of blood samples, while long-term cumulative exposure may be identified through analysis of hair, bone, teeth or nail. Lead is also known to inhibit the activity of an enzyme (ALAD), which can be thus used as an indirect biomarker to estimate human exposure to lead.[27] Cotinine is an example of a urinary biomarker which has been used to validate personal monitoring and questionnaire data associated with exposure to environmental tobacco smoke (ETS) in indoor environments.[25,28] Another non-invasive biomonitoring method is the use of a fingertip pulse oximeter to determine carboxyhemoglobin (*i.e.* hemoglobin combined with CO) levels in blood for detecting personal exposure to CO.

3.2 Exposure Modelling

Exposure modelling is an indirect method for assessing human exposure to air pollution based on a range of indicators, such as proximity to the source, road traffic density, indoor and outdoor pollutant concentrations, *etc.* Although the complexity of exposure models varies widely, they usually involve some kind of mathematical simulations that require input information (*e.g.* distance from source and meteorological data) as well as independent exposure monitoring datasets for validation. Models are particularly useful for assessing exposure in areas where measurements are not available, as well as for testing exposure scenarios and mitigation strategies, reconstructing past exposures and providing exposure predictions for epidemiological studies.[29,30]

Based on their underlying assumptions, exposure models can be broadly classified into the following categories: (A) source-oriented models (including proximity and buffer zone models, dispersion and photochemical air quality models); (B) geostatistical interpolation and land-use regression models; (C) receptor-oriented models (including chemical mass balance and statistical factor analysis models); and (D) time–activity (or micro-environmental) models.

Source-oriented models generally rely on information associated with the pollution source, for example proximity to an industrial source, emission rates, stack height, *etc.*, to predict pollutant concentrations at outdoor locations. Geostatistical interpolation and land-use regression models use statistical techniques to interpolate air quality data obtained from a network of monitoring sites. Receptor-oriented (or *emission-oriented*) models analyse air quality data obtained from one or more receptor locations to identify contributing emission sources. Finally, time–activity models use a combination of air quality measurements, dispersion model simulations and questionnaires to reconstruct the personal exposure of individuals or population subgroups. It should be kept in mind that there are different ways of classifying models, and some exposure models (sometimes called *hybrid*) may combine more than one of the above approaches.

3.2.1 Proximity and Buffer Zone Models

Proximity models are based on the assumption that pollutant concentrations decay with distance from the emission source. This distance (*e.g.* between an industrial plant and a house) can be used as a surrogate of personal exposure. For simplicity, population counts can be allocated to discrete *exposure zones* (or *buffer zones*) determined by distance from major emission sources, such as heavily trafficked streets, motorways, industrial sources, *etc.* The simplest approach is to use circular buffer zones to characterise exposure levels around individual point sources (*e.g.* a power plant). In this case, it is assumed that all residents living within each zone are exposed to uniform air pollutant concentrations. The size of the exposed population can be determined by summing the population counts in each postcode within the zone obtained from census data (Figure 2). The population-weighted average exposure (E_P) in the affected area can be then calculated using the relationship:

$$E_P = \frac{\sum_i C_i \times P_i}{\sum_i P_i} \qquad (1)$$

where C_i is the average concentration in zone i, and P_i the size of the population subgroup exposed in zone i. Empirical indoor–outdoor relationships may also be used to refine the exposure estimates by assigning an average I : O concentration ratio and percentage of time spent indoors to all residents living within certain exposure zone. Although there has been certain criticism on using discrete exposure zones such as road buffers,[29,31] this exposure modelling approach may be preferable in cases where detailed input data (emission rates, meteorological information, *etc.*) and air quality monitoring data for model validation are not available. Another obvious limitation of this method is the assumption that ambient pollutant concentrations at the home address are representative of the total personal exposure of residents who may spend a large proportion of their time at work or commuting. It should be noted that

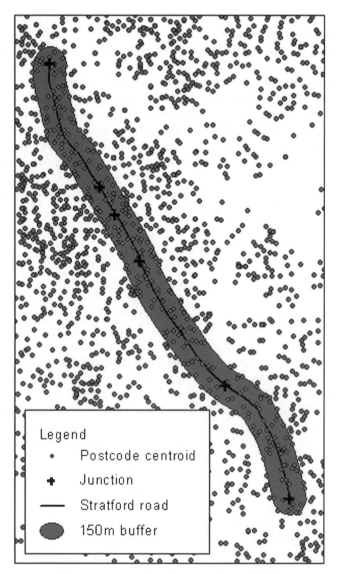

Figure 2 Exposure buffer zones and postcode centroids around a commuter route in Birmingham UK.[10]

differences between day- and night-time counts of residents are not usually reflected in population census data.

3.2.2 Dispersion and Photochemical Air Quality Models

Dispersion models are source-oriented simulation tools commonly used in population exposure studies. One of the most widely used approaches to dispersion modelling is based on Gaussian plume theory, which assumes that dispersion in the horizontal and vertical direction takes the form of a normal probability distribution with the maximum concentration at the centre of the plume (Figure 3).

This modelling methodology can be applied to continuous emissions from stationary (*e.g.* industry and domestic heating) and mobile sources (*e.g.* road transport). Emission sources are simulated point (*e.g.* industrial stack of certain diameter and height), line (*e.g.* motorway), area (*e.g.* landfill) or volume sources (*e.g.* fugitive emissions around buildings). The distribution of pollutants both vertically and horizontally is a function of downwind distance from the source, but also depends on the stability of the atmosphere, the wind speed and direction, the height of the release and other factors. The main advantage of Gaussian plume models is that they are fast and relatively easy to use, require limited input data, and can include urban topography features such as buildings and hills. On the other hand, they are not reliable for very low wind speed ($< 1\,\mathrm{m\,s^{-1}}$) or very variable wind direction or at very long distances from the emission source ($> 10\,\mathrm{km}$). Gaussian-type models (usually called *Gaussian puff models*) may also be used to simulate releases from instantaneous emission sources, such as explosive accidents, occurring over short time periods. Another advantage of Gaussian plume models is that they can easily be linked with GIS, providing overlapping air pollution and population density maps. An example of a Gaussian dispersion model interfaced with GIS is the AirGIS modelling system developed at the Danish National Environmental Research Institute. AirGIS can estimate exposure levels at residence addresses by combining digital maps of urban areas and road traffic information with the Operational Street Pollution Model (OSPM). This system has been applied

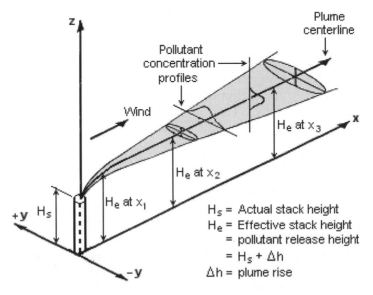

Figure 3 Visualization of a buoyant Gaussian air pollutant dispersion plume.

to human exposure studies, air pollution epidemiological studies, and urban air quality management studies in Denmark and elsewhere.[32,33]

Two more physically realistic, but numerically complicated and computationally expensive, air quality modelling methods are the Eulerian and Lagrangian approaches. Eulerian and Lagrangian models can provide realistic simulations of the atmospheric transport and mixing of air pollutants at regional, urban and neighbourhood scale.[34] Both methodologies can support photochemical models of varying complexity able to simulate chemical transformation of reactive species (*e.g.* VOC, NO_x and ozone) in the atmosphere. Computational Fluid Dynamics (CFD) models, based on the Eulerian approach and structured around numerical algorithms, have used fine grids and advanced turbulence treatment schemes to simulate air flow and pollutant dispersion in street canyons, around and inside buildings, and within the breathing zone of individuals.[35]

Air quality models require a range of input data, including meteorological and local topography data, and emission rates. The contribution of the background needs to be added to account for the fraction of the pollutant that is not emitted from the simulated sources. Usually the urban background concentration is obtained from fixed air quality monitoring sites which are not directly affected by local emission sources. In more complex modelling systems, the urban background concentration can be the obtained from a larger scale (*e.g.* regional) air quality model. It should be stressed that the acquisition and pre-processing of input data is an important part of the modelling study, since the performance of a model depends on the quality of the inputs.[4] Air quality monitoring data need to be used for the validation of the model during its development stages, and later for verification in different applications. Data assimilation (or *blending*) methodologies that combine modelled pollutant concentration fields with observations have also been used in population exposure studies.[36]

3.2.3 Geostatistical Interpolation and Land-Use Regression Models

A range of geostatistical interpolation modelling techniques can be used in exposure assessment studies. These include interpolation models, such as kriging and inverse distance weighing, which generate pollution maps using measurements from monitoring locations distributed over an urban area.[29] Although potentially more reliable than simple proximity models, interpolation methodologies have the limitation of assuming spatially isotropic variation of air pollution (*i.e.* meteorological conditions, traffic density and building effects are not taken into account). Interpolation models become more reliable when air quality data from a dense network of monitoring sites are available (*e.g.* NO_2 diffusion sampling networks).

A more complex statistical technique for producing air pollutant concentration maps based on monitoring data from urban sites is land-use regression modelling. Apart from using air quality measurements and distances from monitoring sites, this method also relies on a range of independent variables, such as land-use type, road traffic density and meteorology. Although increasingly used in exposure studies,[37] land-use regression models are site specific, which means that they cannot be readily extrapolated to other areas with very different land-use characteristics.[29]

3.2.4 Receptor-Oriented Models

Receptor-oriented models are used to estimate the contribution of potential emission sources from pollutant concentrations measured at one or more receptor locations. There are two main types of such models: the Chemical Mass Balance (CMB) models and the statistical factor analysis models. CMB modelling requires source profiles, *e.g.* speciated PM or VOC, and can be carried out for one or more air quality samples. Factor analysis models, such as Principal Component Analysis (PCA) and Positive Matrix Factorization (PMF) models, are based on statistical multivariate methods which require a large amount of measurements obtained over a relatively long period of time (but they do not require speciation). Depending on the data available, both methods can be used to identify the relative source contributions at the location of a receptor without using emission rates. However, they can only identify generic sources (*e.g.* road transport or energy generation) – not individual sources (*e.g.* a specific motorway or power plant). Receptor-oriented modelling has been extensively used to estimate the contribution of different sources to mixtures of pollutants, such as PM_{10} and VOC, in urban environments.[38-40] The method is now increasingly used in personal exposure studies.[41]

3.2.5 Time–Activity (or Micro-Environmental) Modelling

The models described so far estimate pollutant concentrations in ambient air rather than personal exposure. They have been extensively used in population exposure studies assuming that outdoor pollution at residence address is a reliable surrogate of personal exposure. This convenient assumption may not be valid for commuters, passive smokers, individuals exposed to high levels of air pollution in the workplace or people living in poorly ventilated houses with strong indoor sources (*e.g.* gas cooking).

An alternative approach is time–activity (or micro-environmental) modelling which is based on the assumption that the exposure of individuals (or groups) to air pollution can be reconstructed by summing the time-weighted concentrations of the pollutant of interest in different *micro-environments* where people spend most of their time. This can be expressed mathematically with the following relationship:

$$E_H = \frac{\sum_j C_j \times T_j}{\sum_j T_j} \qquad (2)$$

where C_j is the average concentration in micro-environment j, and T_j the time spent in micro-environment j. These micro-environments may include indoor, outdoor and in-vehicle (or *transient*) locations. It is practically impossible to account for all micro-environments as people move continuously in their daily life, but it is important to include in the calculations all places where individuals spend a substantial proportion of their time (home, work-place or school), as well as those micro-environments and activities which are likely to contribute to peak exposures (*e.g.* car driving, filling station, workshop).

The input concentrations used in micro-environmental models can be pooled values of measurements and/or model simulations. For example, the OSPM dispersion model and the INTAIR compartmental model have been used to estimate roadside and indoor pollutant concentrations, respectively, in expo-sure studies in the UK.[10,42] Time–activity profiles can be constructed from specially designed diaries and questionnaires. Population-oriented micro-environmental models may use census population data and average time–activity patterns. For example, the Stochastic Human Exposure and Dose Simulation for Particulate Matter (SHEDS-PM) model estimates the exposure of a statistical individual using the US Environmental Protection Agency (EPA) Consolidated Human Activity Database.[43] In SHEDS-PM, input values for each individual are randomly sampled from probability distributions that represent the variability in the exposure factors.

Micro-environmental models have proved to be reliable in predicting per-sonal exposure to regulated pollutants, such as CO and NO_2. The same is true for PM_{10} when a *personal cloud* (*i.e.* excess personal exposure) increment is added.[44] Micro-environmental models have also been used to reconstruct personal exposures to a wide range of VOC and PAH in urban, suburban and rural areas in the UK.[25] More complex, dynamic GIS-based modelling systems, integrating emission, dispersion and time–activity submodels, have also been developed to assess journey-time exposure to traffic-related air pollution.[45]

3.3   Modelling Uncertainty

An important aspect of exposure modelling is the assessment of uncertainty associated with the structure (*structural uncertainty*) and input parameters of the model (*parametric uncertainty*). Qualitative and quantitative approaches can be used to estimate model uncertainty in exposure studies. For example, in the US National Human Exposure Assessment Survey (NHEXAS), expert sub-jective judgment elicitation techniques were used to characterise the magnitude of uncertainty in environmental exposure to benzene.[46] Deterministic sensitivity analysis and Monte Carlo analysis have also been used to obtain quantitative estimates of uncertainty associated with model input parameters.[10,43]

Sensitivity analysis helps to identify the model parameters that carry more weight in the exposure assessment calculations. A sensitivity analysis can be performed by examining the effect of a range of plausible values of one input parameter at a time on the calculated exposure levels. This gives a rough estimate of the parametric uncertainty associated with the calculations, and indicates where emphasis should be placed in more detailed assessments.

Monte Carlo analysis enables a probabilistic evaluation of the output of the model for many sets of combinations of the input parameters. Input values are sampled randomly from probability distributions assigned to each one of the uncertain variables. This method can be applied to a wide range of input parameters such as outdoor and indoor concentrations, time spent in different micro-environments, resident population within specific exposure zones, *etc.* The probability distributions can be either obtained from empirical evidence or elicited from expert judgements.

4   Evidence from Personal Exposure Surveys in Urban Areas

Several human exposure surveys have been reported in the literature, mainly focusing on VOC (such as benzene), $PM_{2.5}$, CO and NO_2 in indoor residential, workplace, school, transport, and outdoor micro-environments. Exposure to other groups of pollutants, such as PAH and carbonyls (including formaldehyde), has been relatively less studied. Although most surveys have used healthy, non-smoking, working age volunteers, certain studies have included more vulnerable population subgroups (children and elderly) as well as active and passive smokers. A summary of selected exposure studies from North America and Europe are presented in sections 4.1 and 4.2.

4.1   North American Studies

The TEAM (Total Exposure Assessment Measurement) study was an extensive survey of personal exposure to VOC in air (and drinking water) carried out in several US cities, including urban, suburban and rural areas in industrial and semi-industrial settings. Cancer risks associated with personal exposure to these chemicals were calculated for population subgroups, including smokers and non-smokers, and extrapolated to the entire US population.[47] Despite the large uncertainties in exposure–response relationships, this pioneering study quantified cancer risks associated with indoor VOC, highlighting their importance by comparing them with other indoor environmental risks such as passive smoking and radon exposure.

The more recent RIOPA study (Relationship of Indoor, Outdoor and Personal Air) carried out in three US cities has focused on personal exposure to air toxics. This included measurements of VOC, carbonyls, fine particle mass ($PM_{2.5}$) and component species (organic and elemental carbon, and PAH) in approximately 100 non-smoking homes in each of the three cities. Passive and active sampling devices were used to collect simultaneously 48 hour integrated outdoor, indoor and personal air samples. Exposure questionnaires were also used to identify influencing personal activities, as well as home and neighbourhood characteristics. In addition, air exchange rates were experimentally characterised in each home.[48] The study results showed that formaldehyde and acetaldehyde had the strongest indoor source strengths.

I:O ratios for the targeted pollutants were reported and personal exposure to air pollutants was estimated using residential measurements and time–activity data.[49,50] The similar DEARS study (Detroit Exposure and Aerosol Research Study) has assessed the impacts of local industrial and mobile sources on human exposures to air pollutants in Detroit, Michigan.[51] In this study, daily integrated personal exposure, residential indoor and outdoor measurements of VOC and PM constituents were collected in order to assess micro-environmental air quality and evaluate personal exposure models.

Although not an exposure survey, it is worth mentioning the National Air Toxics Assessment (NATA) carried out by the US EPA. This modelling study estimated population exposure to 177 air pollutants (a subset of the 187 air toxics included in the US Clean Air Act, in addition to diesel PM). The exposure assessment was carried out in three steps: (a) initially atmospheric emissions from outdoor sources were estimated, (b) then outdoor concentrations were calculated using the ASPEN dispersion model, and (c) population exposures across the USA were estimated with the HAPEM model using calculated outdoor, indoor and vehicle concentrations, in combination with a wide range of representative time–activity patterns. Finally, the exposure estimates were used to characterise population risks from inhalation of air toxics using toxicity benchmarks.[52]

4.2   European Studies

EXPOLIS was one of the most extensive experimental surveys in Europe focusing on population exposure to a range of air pollutants in large and medium-size cities.[53] Simultaneous personal exposures and micro-environmental measurements (home indoor/outdoor and workplace indoor) of $PM_{2.5}$, VOC, CO and NO_2 (in some of the cities) were taken over 48 hour periods. Personal exposure to $PM_{2.5}$ concentrations was found to be typically higher than respective residential indoor/outdoor and workplace indoor concentrations. Presence of smoking was the strongest predictor of personal exposure to $PM_{2.5}$. Residence location (urban *vs.* suburban) and traffic density in the nearest street from home were among the predictors of $PM_{2.5}$ for individuals not exposed to ETS.[54] Peak personal exposure to CO was associated with smoking, cooking and transport activities.[55] Smoking and road traffic were also identified as the dominant sources of VOC exposure, with ETS-exposed individuals having elevated personal exposure to benzene, toluene, styrene and xylenes.[56] It was demonstrated that time-weighted residential indoor and workplace micro-environmental concentrations could satisfactorily predict personal exposure of non ETS-exposed individuals to VOC that are not strongly associated with road traffic. In certain urban environments, seasonal activities and sources were found to dramatically influence micro-environmental and personal VOC exposures.

The more recent MATCH study (Measurement and modelling of exposure to Air Toxic Concentrations for Health effect studies and verification by biomarker) has investigated the magnitude and range of personal exposures to air toxics in the UK. That included benzene, toluene, xylenes, 1,3-butadiene, benzo[*a*]pyrene, and other VOC and PAH in urban, suburban and rural environments. Personal exposure measurements involving 100 volunteers were carried out using pumped samplers. Lifestyle questionnaires, time–activity diaries and urine samples for ETS biomarker analysis were also collected. In addition to the personal exposure measurements, VOC and PAH samples were collected in homes, offices, street canyons, vehicles, and other indoor and outdoor micro-environments at different times of the day and in different seasons, showing generally higher pollutant concentrations during peak-traffic hours in winter. It was demonstrated that indoor residential concentrations and ETS exposure were the dominant contributors to total personal exposure. The urinary ETS biomarkers correlated strongly with high molecular weight PAH in the personal exposure samples.[25]

Within the framework of another recent European study, AIRMEX (Indoor Air Monitoring and Exposure Assessment Study), exposure measurements were conducted in schools and other public buildings in several south and central European cities. The objective of the study was to estimate I : O pollutant relationships and personal exposure to selected VOC, including aromatics, carbonyls and terpenoids.[57] In agreement with earlier surveys, results from this study indicated that indoor concentrations were significantly higher than outdoor concentrations, and that personal exposures were in most cases substantially higher than both indoor and outdoor pollution levels.

5 Role of Exposure Assessment in Regulatory Control of Air Pollution

For practical reasons, non-occupational air quality regulations have typically focused on ambient concentrations of air pollutants rather than population exposure. This approach is based on the simplifying assumption that only one or a few fixed sampling points can sufficiently represent population exposure in cities, which reflects the design of earlier air quality epidemiology studies. According to current EU legislation, ambient air quality monitoring should be carried out where the highest population exposure to air pollution is likely to occur, but measurements should be representative of the air quality in the area surrounding the receptors.

The common legislative approach focuses on eliminating pollution hotspots by imposing ambient air quality limit values. These limit values ensure that the whole population experience acceptable air quality levels, which reflects the concept of environmental justice. Apart from conventional limit values for key pollutants, an exposure reduction objective has recently been adopted in the UK. This involves a 15% reduction in $PM_{2.5}$ concentrations in urban areas, which is expected to progressively reduce population exposures in the most densely populated parts of the country. The relative benefits of the two approaches, limit values and exposure reduction, for a non-threshold pollutant[58] are discussed in the chapter by M. Williams.

Although indoor air pollution has a proportionally larger impact on personal exposure than outdoor concentrations, indoor air quality is only indirectly controlled through regulations for construction materials, energy performance of buildings, gas and heating appliances, consumer products and chemicals safety. It has often been argued that regulating indoor residential environments impinges on personal freedom. Such regulations would be also very difficult to implement and enforce effectively. Nevertheless, there have been examples of successfully regulating certain environmental hazards, such as radon, in houses. The UK Committee on the Medical Effects of Air Pollutants[59] has recommended guideline values for five pollutants (NO_2, CO, formaldehyde, benzene and benzo[*a*]pyrene) in indoor air. Currently, the World Health Organisation[60] is developing indoor air quality guidelines focusing on a similar list of chemicals.

6 Conclusions

Human exposure to air pollutants can be assessed at personal and population level using a wide range of monitoring and modelling techniques. All these techniques have certain advantages and limitations associated with their accuracy, temporal and spatial resolution, portability of equipment and cost that make them more or less suitable for different applications. Most epidemiological and risk assessment studies have so far been based on ambient air pollutant concentrations rather than population exposure estimates. Although this practice is widely accepted, it does raise some concern about potential exposure misclassification. Some of the shortcomings of conventional air pollution exposure assessments may be avoided by adopting the more advanced techniques (*e.g.* personal exposure monitoring and dynamic time–space modelling) presented in this chapter. However, an optimum balance between practicality and complexity needs to be struck on a case by case basis, depending on study aims, expertise and resources available.

Strategies to reduce population exposure to air pollutants would generally benefit from a combination of ambient air quality limit values focusing on pollution hotspots, exposure reduction objectives for non-threshold pollutants, and a range of vehicle emission standards, building standards, consumer products and chemicals safety regulations targeting air pollutants in indoor and outdoor environments.

References

1. O. Hertel, F. A. A. M. De Leeuw, O. Raaschou-Nielsen, S. S. Jensen, D. Gee, O. Herbarth, S. Pryor, F. Palmgren and E. Olsen, *Pure Appl. Chem.*, 2001, **73**(6), 933–958.
2. T. E. McKone, P. B. Ryan and H. Özkaynak, *J. Exposure Sci. Environ. Epidemiol.*, 2009, **19**(1), 30–44.
3. O. Raaschou-Nielsen, J. H. Olsen, O. Hertel, R. Berkowicz, H. Skov, A. M. Hansen and C. Lohse, *Sci. Total Environ.*, 1996, **189**(190), 51–55.

4. S. Vardoulakis, B. E. A. Fisher, K. Pericleous and N. Gonzalez-Flesca, *Atmos. Environ.*, 2003, **37**(2), 155–182.
5. B. Croxford and A. Penn, *Sci. Total Environ.*, 1996, **189**(190), 3–9.
6. C. Monn, V. Carabias, M. Junker, R. Waeber, M. Karrer and H. U. Wanner, *Atmosph. Environ.*, 1997, **31**(15), 2243–2247.
7. J. I. Levy, D. H. Bennett, S. J. Melly and J. D. Spengler, *J. Exposure Anal. Environ. Epidemiol.*, 2003, **13**(5), 364–371.
8. E. P. Weijers, A. Y. Khlystov, G. P. A. Kos and J. W. Erisman, *Atmosph. Environ.*, 2004, **38**(19), 2993–3002.
9. S. Vardoulakis, N. Gonzalez-Flesca, B. E. A. Fisher and K. Pericleous, *Atmos. Environ.*, 2005, **39**(15), 2725–2736.
10. S. Vardoulakis, Z. Chalabi, T. Fletcher, C. Grundy and G. S. Leonardi, *Sci. Total Environ.*, 2008, **394**(2–3), 244–251.
11. S. Vardoulakis, N. Gonzalez-Flesca and B. E. A. Fisher, *Atmos. Environ.*, 2002, **36**(6), 1025–1039.
12. J. I. Levy, *J. Air Waste Manage. Assoc.*, 1998, **48**(6), 553–560.
13. P. Pérez Ballesta, R. A. Field, R. Connolly, N. Cao, A. B. Caracena and E. De Saeger, *Atmos. Environ.*, 2006, **40**(18), 3355–3366.
14. H. K. Lai, M. J. Jantunen, N. Kunzli, E. Kulinskaya, R. Colvile and M. Nieuwenhuijsen, *Atmos. Environ.*, 2007, **41**(39), 9128–9135.
15. J. F. Zhang and K. R. Smith, *Br. Med. Bull.*, 2003, **68**, 209–225.
16. C. L. Townsend and R. L. Maynard, *Occup. Environ. Med.*, 2002, **59**(10), 708–711.
17. B. C. Singer, A. T. Hodgson, T. Hotchi and J. J. Kim, *Atmos. Environ.*, 2004, **38**(3), 393–403.
18. L. A. Wallace, E. D. Pellizzari, T. D. Hartwell, V. Davis, L. C. Michael and R. W. Whitmore, *Environ. Res.*, 1989, **50**(1), 37–55.
19. K. Stevenson, T. Bush and D. Mooney, *Atmos. Environ.*, 2001, **35**(2), 281–287.
20. B. Croxford and A. Penn, *Atmos. Environ.*, 1998, **32**(6), 1049–1057.
21. A. Scaperdas and R. N. Colvile, *Atmos. Environ.*, 1999, **33**, 661–674.
22. F. S. Violante, A. Barbieri, S. Curti, G. Sanguinetti, F. Graziosi and S. Mattioli, *Chemosphere*, 2006, **64**(10), 1722–1729.
23. J. Sunyer, D. Jarvis, T. Gotschi, R. Garcia-Esteban, B. Jacquemin, I. Aguilera, U. Ackerman, R. De Marco, B. Forsberg, T. Gislason, J. Heinrich, D. Norback, S. Villani and N. Kunzli, *Occup. Environ. Med.*, 2006, **63**(12), 836–843.
24. S. Kaur, M. Nieuwenhuijsen and R. Colvile, *Atmos. Environ.*, 2005, **39**(20), 3629–3641.
25. R. M. Harrison, J. M. Delgado Saborit, S. Baker, N. Aquilina, C. Meddings, S. Harrad, I. Matthews, B. Armstrong, S. Vardoulakis and R. Anderson, *Measurement and Modelling of Exposure to Air Toxic Concentrations for Health Effect Studies and Verification by Biomarker (MATCH Study)*, University of Birmingham, UK, 2008.
26. M. Nieuwenhuijsen, *Exposure Assessment in Occupational and Environmental Epidemiology*, Oxford University Press, Oxford, 2003.

27. F. Barbosa, J. E. Tanus-Santos, R. F. Gerlach and P. J. Parsons, *Environ. Health Persp.*, 2005, **113**(12), 1669–1674.

28. A. H. Wu, in *Topics in Environmental Epidemiology, K. Steenland,* Oxford University Press, Oxford, 1997, pp. 200–226.

29. M. Jerrett, A. Arain, P. Kanaroglou, B. Beckerman, D. Potoglou, T. Sahsuvaroglu, J. Morrison and C. Giovis, *J. Exposure Anal. Environ. Epidemiol.*, 2005, **15**(2), 185–204.

30. M. Nieuwenhuijsen, D. Paustenbach and R. Duarte-Davidson, *Environ. Int.*, 2006, **32**(8), 996–1009.

31. J. Molitor, M. Jerrett, C. C. Chang, N. T. Molitor, J. Gauderman, K. Berhane, R. McConnell, F. Lurmann, J. Wu, A. Winer and D. Thomas, *Environ. Health Persp.*, 2007, **115**(8), 1147–1153.

32. S. S. Jensen, *J. Hazard. Mater.*, 1998, **61**(1–3), 385–392.

33. O. Hertel, S. S. Jensen, H. V. Andersen, F. Palmgren, P. Wahlin, H. Skov H, I. V. Nielsen, M. Sorensen, S. Loft and O. Raaschou-Nielsen, *Pure Appl. Chem.*, 2001, **73**(1), 137–145.

34. C. Borrego, O. Tchepel, A. M. Costa, H. Martins, J. Ferreira and A. I. Miranda, *Atmos. Environ.*, 2006, **40**(37), 7205–7214.

35. A. J. Gadgil, C. Lobscheid, M. O. Abadie and E. U. Finlayson, *Atmos. Environ.*, 2003, **37**(39–40), 5577–5586.

36. W. L. Physick, M. E. Cope, S. Lee and P. J. Hurley, *J. Exposure Sci. Environ. Epidemiol.*, 2007, **17**(1), 76–83.

37. G. Hoek, R. Beelen, K. De Hoogh, D. Vienneau, J. Gulliver, P. Fischer and D. Briggs, *Atmos. Environ.*, 2008, **42**(33), 7561–7578.

38. R. M. Harrison, A. M. Jones and R. G. Lawrence, *Atmos. Environ.*, 2003, **37**(35), 4927–4933.

39. S. Vardoulakis and P. Kassomenos, *Atmos. Environ.*, 2008, **42**(17), 3949–3963.

40. Y. Song, W. Dai, M. Shao, Y. Liu, S. Lu, W. Kuster and P. Goldan, *Environ. Pollution.*, 2008, **156**, 174–183.

41. M. J. Anderson, S. L. Miller and J. B. Milford, *J. Exposure Anal. Environ. Epidemiol.*, 2001, **11**(4), 295–307.

42. C. Dimitroulopoulou, M. R. Ashmore, M. A. Byrne and R. P. Kinnersley, *Atmos. Environ.*, 2001, **35**(2), 269–279.

43. J. M. Burke, M. J. Zufall and H. Ozkaynak, *J. Exposure Anal. Environ. Epidemiol.*, 2001, **11**(6), 470–489.

44. R. M. Harrison, C. A. Thornton, R. G. Lawrence, D. Mark, R. P. Kinnersley and J. G. Ayres, *Occup. Environ. Med.*, 2002, **59**(10), 671–679.

45. J. Gulliver and D. J. Briggs, *Environ. Res.*, 2005, **97**(1), 10–25.

46. K. D. Walker, J. S. Evans and D. Macintosh, *J. Exposure Anal. Environ. Epidemiol.*, 2001, **11**(4), 308–322.

47. L. A. Wallace, *Environ. Health Persp.*, 1991, **95**, 7–13.

48. C. P. Weisel, J. F. Zhang, B. J. Turpin, M. T. Morandi, S. Colome, T. H. Stock, D. M. Spektor, L. Korn, A. Winer, S. Alimokhtari, J. Kwon, K. Mohan, R. Harrington, R. Giovanetti, W. Cui, M. Afshar, S. Maberti and D. Shendell Relationship of Iindoor, *J. Exposure Anal. Environ. Epidemiol.*, 2005, **15**(2), 123–137.

49. W. Liu, J. Zhang, L. Zhang, B. J. Turpin, C. P. Welsel, M. T. Morandi, T. H. Stock, S. Colome and L. R. Korn, *Atmos. Environ.*, 2006, **40**(12), 2202–2214.

50. W. Liu, J. J. Zhang, L. R. Korn, L. Zhang, C. P. Weisel, B. Turpin, M. Morandi, T. Stock and S. Colome, *Atmos. Environ.*, 2007, **41**(25), 5280–5288.

51. R. Williams, A. Vette, J. Burke, G. Norris, K. Wesson, M. Strum, T. Fox, R. Duvall and T. Watkins, *Air Pollut. Model. Appl. XIX*, 2008, 727–728.

52. USEPA. *The National-scale Air Toxics Assessment.* 2008.

53. M. J. Jantunen, O. Hanninen, K. Katsouyanni, H. Knoppel, N. Kuenzli, E. Lebret, M. Maroni, K. Saarela, R. Sram and D. Zmirou, *J. Exposure Anal. Environ. Epidemiol.*, 1998, **8**(4), 495–518.

54. K. J. Koistinen, O. Hanninen, T. Rotko, R. D. Edwards, D. Moschandreas and M. J. Jantunen, *Atmos. Environ.*, 2001, **35**(14), 2473–2481.

55. H. K. Lai, M. Kendall, H. Ferrier, I. Lindup, S. Alm, O. Hanninen, M. Jantunen, P. Mathys, R. Colvile, M. R. Ashmore, P. Cullinan and M. J. Nieuwenhuijsen, *Atmos. Environ.*, 2004, **38**(37), 6399–6410.

56. R. D. Edwards, J. Jurvelin, K. Saarela and M. Jantunen, *Atmos. Environ.*, 2001, **35**, 4531–4543.

57. D. Kotzias, *Exp. Toxicol. Pathol.*, 2005, **57**, 5–7.

58. DEFRA, *The Air Quality Strategy for England, Scotland, Wales and Northern Ireland (Volume 2)*, Department for Environment, Food and Rural Affairs in partnership with the Scottish Executive, Welsh Assembly Government and Department of the Environment North Ireland, UK, 2007.

59. COMEAP, *Recommended indoor air quality guidelines for homes*, Committee on the Medical Effects of Air Pollutants, UK Dept. Health, 2004.

60. WHO, *Development of WHO Guidelines for Indoor Air Quality*, Report on a Working Group Meeting in Bonn, Germany, 23–24 Oct. 2006, WHO, Copenhagen, Denmark, 2006.

61. N. A. H. Janssen, P. H. N. Van Vliet, F. Aarts, H. Harssema and B. Brunekreef, *Atmos. Environ.*, 2001, **35**(22), 3875–3884.

62. P. Schneider, I. Gebefugi, K. Richter, G. Wolke, J. Schnelle, H. E. Wichmann and J. Heinrich, *Sci. Total Environ.*, 2001, **267**(1–3), 41–51.

63. H. Fromme, T. Lahrz, A. Piloty, H. Gebhardt, A. Oddoy and H. Ruden, *Sci. Total Environ.*, 2004, **326**(1–3), 143–149.

Health Effects of Urban Pollution

ROBERT L. MAYNARD

ABSTRACT

More research on the effects of air pollutants on health is probably being done today than at any time in history. This is remarkable in view of the fact that, in many developed countries, concentrations of many air pollutants are now low in comparison with earlier periods. The position is much more alarming in the developing world: there pollutant concentrations are rising and effects on health are increasing. Present concerns about even low concentrations of air pollutants have been fuelled by developments in epidemiological techniques: time-series analysis has played a large part in demonstrating that even for common and non-carcinogenic air pollutants such as ozone and sulfur dioxide there may be no threshold of effect - at least, not at a population level. Findings from these studies are discussed in this chapter. Another remarkable advance has been the realisation that long-term exposure to the ambient aerosol increases the likelihood, at all adult ages, of death from cardiovascular disease. This effect may be due to an increased rate of development of atheromatous plaques in the coronary arteries with rupture of plaques leading to myocardial infarction. Time-series studies have revealed that even short-term exposure to particles increases the likelihood of cardiovascular "events" and abnormalities in the rhythm of the heart. It has been suggested that the ultrafine component of the ambient aerosol plays a large part in causing these effects. This has, in part, led to the current surge of interest in nano-toxicology. The suggestion remains unproven.

Issues in Environmental Science and Technology, 28
Air Quality in Urban Environments
Edited by R.E. Hester and R.M. Harrison
© Royal Society of Chemistry 2009
Published by the Royal Society of Chemistry, www.rsc.org

1 Introduction

Research on the effects of air pollutants on health has progressed rapidly during the past ten years. Though the number of pollutants of concern has not been added to, our understanding of their impacts on health, if not of their mechanisms of effect, has developed to a point where impacts on health can now be predicted with some confidence. This is, in part, due to the large number of studies reported, in part because of advances in epidemiological methods and lastly, because of developments in meta-analytical techniques. Some surprises have occurred: long-term exposure to particles affects the heart more than the lung; thresholds of effect, even for pollutants such as ozone and sulfur dioxide, seem increasingly unlikely; and ultrafine particles (nanoparticles) may be the most active component of the ambient aerosol. Calculations of the impacts of air pollutants on health have been undertaken in several countries. Such calculations are not, *per se*, difficult and require information on ambient concentrations of pollutants, coefficients linking concentrations of specific pollutants with specific indices of effects on health and, information on background (existing) levels of the indices or end-points. Calculations of effects in the UK suggest that exposure to current concentrations of fine particles ($PM_{2.5}$) imposes an average reduction in life expectancy of about 8 months. Calculations can also be done on the basis of time-series studies. Here the results are expressed in numbers of deaths brought forward and, again in the UK, about 6000 deaths are brought forward each year due to exposure to the ambient aerosol.[1] Similar calculations have been undertaken on a wider basis.[2]

As suggested above, toxicology has lagged behind epidemiology in the air pollution field and we lack, even now, a clear grasp of the mechanisms by which low concentrations of common gases such as ozone and sulfur dioxide affect health. As for particles, though the oxidant hypothesis is well supported, there remain many problems: how active is the so-called coarse fraction of PM_{10}? To what extent does particle composition affect the toxicological properties of particles? If ultrafine particles are the key active component but are not well correlated with PM_{10} (they are not), why does PM_{10} correlate so well with effects on health? In this chapter a few of these issues are explored.

2 Exposure to Air Pollutants

In the air pollution field we know a great deal more about the relationship (or association) between ambient concentrations of pollutants and effects on health than we do about the association between exposure and effects on health. Ambient concentration is easy to measure; personal exposure is not, at least not on a sufficiently widespread basis to allow large scale epidemiological studies. This is important because large scale studies are needed to discover small effects. Small, that is, in terms of coefficients linking concentration and effects, but large in terms of impacts on public health. It is assumed that ambient outdoor concentration is correlated with exposure but people spend most of their lives indoors where concentrations of at least some air pollutants are generally lower than those outdoors. This is the case in developed countries; in developing countries the use of biofuels for cooking and space heating indoors can generate high concentrations of pollutants.[2,3] Despite this, average personal exposure has been found to correlate well with outdoor concentrations on a day-to-day basis.[4] How personal exposure to air pollutants is distributed across the population is less well known: those living close to sources of primary pollutants will be likely to be more exposed than those living at some distance, whereas for secondary pollutants, for example, ozone, the converse may be true. The discrepancy between ambient concentrations and exposure may lead to errors in epidemiological studies: such errors may reduce the size of coefficients linking pollutant concentrations and may make such effects more difficult to detect.[5–7]

3   Epidemiological Approaches

3.1   Time-Series Studies

The classical time-series air pollutant study seeks to link day-to-day variations in ambient concentrations of pollutants with day-to-day changes in counts of health-related effects such as deaths and hospital admissions, making due allowance for other factors that affect such indices and which also vary from day-to-day. These "other" factors are designated as confounding factors: ambient temperature is, perhaps, the most important. Early studies did not involve sophisticated mathematical techniques[8] and led to errors regarding thresholds of effect. More recent studies have shown that for the classical air pollutants (particles, sulfur dioxide, nitrogen dioxide, carbon monoxide, and ozone), no threshold of effect can be identified at a population level. The number of such studies now published is large and meta-analytical methods have been applied to summarise their results. Meta-analysis is a technique that allows the coefficients produced by several, often many, epidemiological studies to be combined. This approach allows a summary coefficient to be identified. Meta-analytical techniques have substantial statistical power and this leads to increased statistical confidence in the derived coefficient. The results obtained by such methods are well shown by "forest plots". Figure 1 is taken from the recent Department of Health report on air pollution and cardiovascular disease.[9] In all the forest plots reproduced in this chapter the *x* axis shows the size of the reported effect, expressed as the percentage change in the effect considered, for example all cause mortality, per specified change in pollutant concentration. In the examples reproduced here the change per $10 \, \mu g$ m^{-3} of pollutant is plotted. The *y* axis lists all the studies included in the meta-analysis. These conventions are illustrated in Figure 1.

The results of 67 studies are shown: for references to the original studies see the Department of Health report.[9] The important feature of the plot is the generally positive association between PM_{10} and cardiovascular mortality: nearly all the central estimates of effect fall to the right of the vertical line showing no change in ill-health. The derived meta-analysis coefficient is both

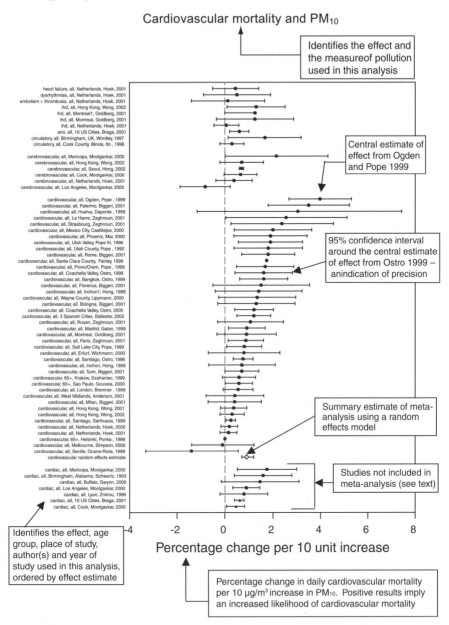

Figure 1 Forest plot showing the effects of PM_{10} on cardiovascular mortality.

positive and statistically significant. It will be noted that there is considerable heterogeneity in the results of the individual studies. This is interesting and may shed light on components of PM_{10} with effects on health. It might also reveal publication bias! Figure 2 shows a similar forest plot for nitrogen dioxide.

112   *Robert L. Maynard*

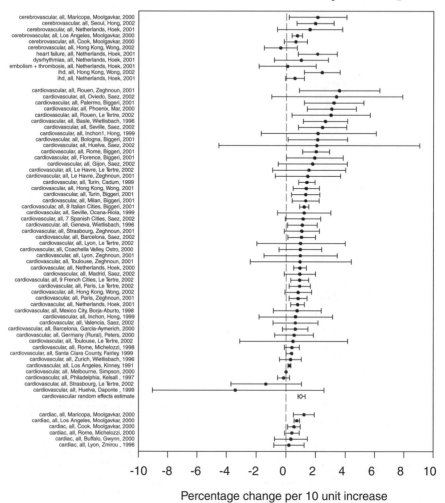

Figure 2 Forest plot showing cardiovascular mortality and daily NO₂ concentrations.

It looks remarkably similar to Figure 1 – is this due to the close correlation between nitrogen dioxide and particle concentrations in urban areas or does nitrogen dioxide have its own effects? This remains an unanswered question. At present, the association with particle concentrations is accepted as more likely to be causal than the association with nitrogen dioxide, but the possibility that nitrogen dioxide could act as an effect modifier remains plausible. The term "effect modifier" may need some explanation: it is sometimes confused with the term "confounding factor". A confounding factor is some factor which varies with the pollutant under examination and which also affects health.

Temperature, for example, is a confounding factor: an increase in temperature above about 18 degrees Centigrade leads to an increase in daily deaths as does an increase in concentrations of ozone. One would need to allow for the effects of temperature in studying the effects of ozone. But an increase in temperature does not, *per se*, increase the effects of ozone. Nitrogen dioxide, on the other hand, does seem to increase the effects of particles: it acts as a modifier of the effects of particles.

3.2 Intervention Studies

Epidemiological studies tend not to be experimental in the sense that they do not involve the study of a deliberate change in what is regarded as the causal agent of ill-health. In the air pollution field, studies of the effects of policies that lead to rapid and significant changes in levels of air pollutants are rare but three have been reported and studied in detail. Banning coal sales in Dublin,[10] reducing the sulfur content of oil used by power plants and road vehicles in Hong Kong[11] and changing transportation patterns before the Olympic Games in Atlanta (USA)[12] all produced improvements in health. Industrial action by workers cannot be regarded as a policy initiative, but studies on the effects of a strike at a steel mill in the Utah Valley[13,14] showed a reduction in admissions of children to hospital for treatment of respiratory diseases during the period when the steel mill was inoperative. It was also shown, uniquely, that the toxicological activity of the ambient aerosol decreased during the strike.

3.3 Cohort Studies

The effects of long-term exposure to particles have been studied by means of well designed cohort studies. In these, the risk of death at all ages has been compared across cities or sub-city areas with different long-term average particle concentrations. In these studies, unlike the time-series studies, the key confounding factors are personal, for example, smoking habits, socio-economic status and occupational history. These do not change from day-to-day (hence they are not relevant in time-series studies) but do vary from person-to-person and so are important in cohort studies. The first study of this kind was reported[15] as part of the Six Cities studies set up by Harvard University in the 1970s. The study was truly prospective and included 8,111 people (aged 25-74) and specially placed pollutant monitors. The second, larger, study[16-18] capitalised on the US Cancer Society cohort and studied 552,138 individuals from 151 US metropolitan areas from which routinely collected data on $PM_{2.5}$ were available from 50 and on sulfate concentrations from all. A recent examination of the findings of these studies[19] has led to the following conclusions:

(i) long-term exposure to a pollution increment of $10\,\mu g\,m^{-3}$ $PM_{2.5}$, is associated with a 6% increase in risk of death, at all ages, from cardiovascular disease;

(ii) the effect on the risk of death from lung diseases, with the exception of lung cancer, is trivial. For lung cancer the increased risk is 14% per $10 \, \mu g \, m^{-3} \, PM_{2.5}$;

(iii) the effect on the risk of death from diseases other than cardiovascular and respiratory is zero. Accidental causes of death, for example from road traffic accidents are excluded from consideration.

These are very simplified conclusions and hardly do justice to the original studies or to the reanalysis of these studies undertaken by the US Health Effects Institute.[20] More recent cohort studies undertaken at a smaller spatial scale[21,22] have reported larger coefficients than the Dockery, Pope studies referred to above. This may be due to a reduction in errors consequent upon equating ambient concentration with exposure across a large area, or perhaps, to the greater toxicity per μg of $PM_{2.5}$ in urban areas where the major source of $PM_{2.5}$ is mainly traffic. Further work in this area is needed.

Having considered the epidemiological approaches that have had most impact on thinking in the past 10 years we now turn to individual air pollutants. Only the classical air pollutants are considered. More detailed accounts can be found[23,24] and are summarised in the World Health Organization Air Quality Guidelines.[25]

4 Particulate Matter

Of all the classical air pollutants, particulate matter is the most intriguing. The bulk composition of particles comprising the ambient aerosol varies from site to site: near the coast sodium chloride is a major component, downwind of coal or oil burning power stations ammonium sulfate is important, in urban areas primary particles formed as organic components of vehicle engine exhaust condense and road dust is resuspended are predominant. Metals are present in all locations: vanadium is a marker for oil combustion, and selenium is a marker for coal. Metal smelters and steel works contribute nickel and iron. More complex organic species, including polycyclic aromatic hydrocarbon (PAH) compounds also occur. Understanding the toxicological properties of this complex mixture is a formidable task and it is remarkable that epidemiological studies have produced such consistent results expressed in terms of the size-specified mass concentration of the aerosol. The size-specification of the mass concentration measurements is important: PM_{10} representing particles of, generally, less than 10 μm aerodynamic diameter and, thus, the fraction of the ambient aerosol that has a high probability of penetrating past the upper airway (nose, mouth, nasopharynx and larynx) and with a size specific probability of being deposited in the lung. Of the particles comprising PM_{10}, a proportion of the larger ones deposit in the conducting airways; the smaller particles are more likely to reach the deeper parts of the lung and a proportion are deposited there. The mass concentration of particles less than 2.5 μm aerodynamic diameter (a subset of PM_{10}), that is, $PM_{2.5}$, is used to represent the category of particles that can, but by no means all do, deposit in the gas-exchange (deep) part of the lung. It is assumed that deposition in this region is more likely to cause toxicological effects than deposition in the conducting airways. Clearance of particles from the deep lung is slower than that from the conducting airways. Also, the delicate epithelium lining the alveoli is generally assumed to be more likely to be damaged than the relatively thick, mucus protected, lining of the conducting airways. Transfer of particles across the alveolar epithelium to the blood and lymphatic systems is more likely than in the conducting airways. Very small particles (<100 nm diameter) deposit especially well in the alveoli. Such particles, described as ultrafine particles or nanoparticles, contribute little to the mass concentration of the ambient aerosol, dominate the number concentration and may be important toxicologically. A unit mass of such particles has a much larger total surface area than the same mass of larger particles and, inevitably, a larger proportion of the atoms or molecules comprising the particles is present at the particle surfaces. All this is likely to lead to greater toxicological activity on a per unit mass basis, though increased activity on a per surface area basis is less likely and, not observed in nanotoxicological studies of titanium dioxide.[26] Some discussion has focused on the role of sulfate in the ambient aerosol. Sulfate ions, *per se*, are not toxicologically active and some have seen the ammonium sulfate component of the ambient aerosol as an essentially inert diluent of more active components. This has policy implications as one means of reducing the mass concentration of the ambient aerosol is to reduce emissions of sulfur dioxide (a precursor of sulfate) by power stations. But, if sulfate is inactive, such a change could not lead to a reduction in the toxicity of the aerosol; on the contrary, as the mass concentration fell (as a result of sulfate concentration falling) the toxicity expressed on a per unit mass basis might well increase. Seeing sulfate as an inactive component may be correct, but it might better be seen as the end product of a reaction sequence that generates toxicologically active components, including hydrogen ions and free metal species. The case for hydrogen ion generation is strong,[27,28] the case for production of soluble metal species (by the reaction of strong acid (H_2SO_4) with insoluble metal oxides) is chemically plausible. Soluble metal species are known to be likely to play a part in free radical formation. Sulfate may then be better regarded as the "process indicator" than as a "source marker".

The mechanisms by which particles affect the cardiovascular system, in fact the heart, are not fully understood. Much emphasis has been placed on ultra-fine particles (<100 nm diameter). Seaton *et al* [29] and Oberdörster *et al* [30] proposed that ultrafine particles might account for the seemingly unlikely effects of very small mass doses of inhaled particles. To the toxicologist it seemed unlikely that a few tens of μg of particles could cause serious injury unless some process independent, as it were, of mass dose was in play. This hypothesis has become established despite rather weak epidemiological support and two theories to explain the effects have been developed. Firstly, it has been suggested that ultrafine particles induce an inflammatory response in the lung – especially, perhaps, in the tissue (interstitium) that lies between the alveolar epithelium and the pulmonary capillary endothelium. This is a narrow space – essentially non-existent at the gas exchange portion of the alveolar septae. Genes controlling the formation of inflammatory cytokines (chemical messengers) have been shown to be activated and changes in clotting factors in the blood have been reported. Other studies[31–37] support these suggestions. Furthermore, research using rabbits predisposed to develop atherosclerotic plaques in their coronary arteries[38] suggests that exposure to particles can accelerate plaque development and destabilisation. Lippmann's group have explored this area in detail in APoE-/- mice exposed to concentrated ambient particles (CAP) in New York and have produced evidence to support these theories.[39–45] Secondly, changes in the autonomic control of the heart have been shown to follow inhalation of particles in man[46] and in animals.[32] The heart is controlled both in terms of its rate (beats per minute) and force of contraction by the autonomic nervous system. This system includes the sympathetic system, commonly associated with "flight and fright" reactions and the perhaps less well known parasympathetic system which slows the heart beat. Important studies by Peters *et al*[47,48] have shown that the frequency of the firing of implanted defibrillator devices is increased, in patients requiring frequent defibrillation, on days with high particle concentrations. A neural reflex is presumably involved, though the afferent arm, that leading from some receptor, in the lung perhaps, to the spinal cord or brain, of the reflex arc is, as yet, imperfectly defined.

These findings represent significant steps forward in our understanding of the effects of particles on health. Such work has led to a reduction in the concentrations of particles recommended by the World Health Organization Air Quality Guidelines. In the latest edition of the guidelines, interim targets as well as guidelines have been proposed. The values are set out in Tables 1 and 2.

5   Nitrogen Dioxide

Given that nitrogen dioxide (NO_2) is, unlike the ambient aerosol, a well defined, single chemical compound, one might expect its effects to be relatively easily understood. Sadly, this is not the case. The main sources of NO_2 in urban areas are vehicles and space heating. Concentrations of NO_2 tend to be rather closely correlated with concentrations of fine particles and this makes separation of the effects of these two pollutants difficult. Some experts regard NO_2 *per se* as toxicologically active in low concentrations; others regard NO_2 merely as a marker for vehicle-generated pollutants: the effects being attributed to particles. This question has been examined by time-series studies which, in general, use single-pollutant models. A meta-analysis of 109 such studies[49] was undertaken: 32 coefficients for NO_2 were included from single-pollutant models and 15 from multi-pollutant models. The "bottom line" result was expressed as:

For single-pollutant models:

A 24 ppb increment in NO_2 is associated with a 2.8% (CI: 2.1–3.5) change in all-cause mortality.

Table 1 Air Quality Guideline and Interim Targets for PM: Annual. (Reproduced with kind permission of the World Health Organization.)

Annual Mean Level	PM_{10} ($\mu g/m^3$)	$PM_{2.5}$ ($\mu g/m^3$)	Basis for the selected level
WHO Interim target 1 (IT-1)	70	35	These levels are estimated to be associated with about 15% higher long-term mortality than at AQG levels
WHO Interim target 2 (IT-2)	50	25	In addition to other health benefits, these levels lower risk of premature mortality by approximately 6% (2-11%) compared to IT-1
WHO Interim target 3 (IT-3)	30	15	In addition to other health benefits, these levels lower risk of premature mortality by approximately another 6% (2-11%) compared to IT-2
WHO air quality guidelines (AQG)	20	10	These are the lowest levels at which total cardiopulmonary and lung cancer mortality have been shown to increase with more than 95% confidence in response to $PM_{2.5}$ in the ACS study (Pope *et al*, 1995; 2002). The use of the $PM_{2.5}$ guideline is preferred.

Table 2 Air Quality Guideline and Interim Targets for PM: 24 hour mean. (Reproduced with kind permission of the World Health Organization.)

24-hour mean level[a]	PM_{10} ($\mu g/m^3$)	$PM_{2.5}$ ($\mu g/m^3$)	Basis for selected level
WHO Interim target 1 (IT-1)	150	75	Based on published risk coefficients from multicentre studies and meta-analyses (about 5% increase in short-term mortality over AQG)
WHO Interim target 2 (IT-2)	100	50	Based on published risk coefficients from multicentre studies and meta-analyses (about 2.5% increase in short-term mortality over AQG)
WHO Interim target 3 (IT-3)[b]	75	37.5	About 1.2% increase in short-term mortality over AQG
WHO air quality guidelines (AQG)	50	25	Based on relationship between 24-hour and annual PM levels

[a]99th percentile (3 days year^{-1}).
[b]For management purposes, based on annual average guideline values, the precise number to be determined on the basis of local frequency distribution of daily means.

For multi-pollutant models:

A 24 ppb increment in NO_2 is associated with a 0.9% (CI: -0.1–2.0) change in all-cause mortality.

It will be noted that the meta-analysis coefficient derived from multi-pollutant models is lower than that derived from single-pollutant models and, in formal statistical terms, lacks significance. This may be due to allowance being made in multi-pollutant models for the effects of other pollutants.

Significant effects have been reported.[50–52] Effects on respiratory admissions to hospital have also been reported.[53–55] In addition to these studies of effects of short-term variations in exposure to NO_2, other workers[56–59] have reported that children growing up in areas with relatively high long-term average concentrations of NO_2 suffer from retarded lung development. These results are supported by work on European adults.[57]

Explaining such effects in toxicological terms has not been successful. Exposure to high concentrations of NO_2 causes damage to the lung and damage to Type I alveolar cells and ciliated airway epithelial cells has been shown, in rats, exposed to 340 ppb (640 μg m^{-3}) NO_2. Long-term exposure to NO_2 has been shown to produce emphysema-like changes in several species.[60] In man, changes in airway resistance (bronchoconstriction) are seen on exposure to 300 ppb (520 μg m^{-3}) NO_2 for 2–2.5 hours[61–63] but the changes tended to be small and of questionable clinical significance. Airway inflammation has been recorded in man on exposure to 1 ppm (1880 μg m^{-3}) NO_2 but not at 600 ppb (1130 μg m^{-3}) – the latter being given as 4 separate 2 hour exposure over 6 days.[64–66] None of these studies provides an explanation for the effects of NO_2 on all-cause mortality reported in time-series studies. The World Health Organization has taken a cautious approach and has recommended the following guidelines[25]

Annual average concentration: 40 μg m^{-3}.
1 hour average concentration: 200 μg m^{-3}.

6   Sulfur Dioxide

Sulfur dioxide (SO_2) is a water soluble gas that is largely absorbed in the upper airways. Despite this, enough can penetrate the bronchi and bronchioles to cause bronchoconstriction, especially in people suffering from asthma. The concentration of SO_2 needed to do this is seldom reached in urban areas of the UK today. Absorbed SO_2 reacts with water to form sulfite and bisulfite ions:

$$SO_2 + H_2O \rightarrow H_2SO_3$$
$$H_2SO_3 + H_2O \rightarrow HSO_3^- + H_3O^+ (pKa = 1.86)$$
$$HSO_3^- + H_2O \rightarrow SO_3^{2-} + H_3O^+ (pKa = 7.2)$$

Bisulfite ions are converted to sulfate ions under the influence, *in vivo*, of a molybdenum containing oxidase enzyme.

Bisulfite ions also react with oxidised glutathione: bisulfite ions are thus a reducing agent and contributed to the reducing activity of coal smoke smog. Sulfurous acid (H_2SO_3) can be converted to sulfuric acid (H_2SO_4) and this may occur at the surface of inhaled particles, especially if catalytically active metal species are present. This may explain the increased effects of SO_2 in the presence of an ultrafine zinc aerosol.[67]

Asthmatic subjects respond with increased airway resistance to inhaling SO_2 at concentrations of 100–400 ppb (286–1144 $\mu g\,m^{-3}$) though significant effects at the lower end of this range are unlikely.[68] Inhaled SO_2 interacts with irritant receptors in the airway wall and this may lead to reflex changes in the control of heart rate.[69] The significance of these changes in subjects with normal myocardial function is unknown. Bronchoconstrictor responses to SO_2 appear, rapidly as is classically the case with irritants: the response is dependent on concentration to a much larger extent than on duration of exposure.

Time-series epidemiological studies show clear associations between all-cause mortality (excluding accidental deaths) and daily variations in SO_2 concentrations. A major European study[70] conducted in 12 cities reported a 2.9% (CI: 2.3–4.5) increase in mortality per 50 $\mu g\,m^{-3}$ increase in SO_2 concentration. Similar findings have been reported elsewhere.[71,72] The large US NMMAPS Study[50,51] reported a smaller coefficient: 1.1% (CI: 0.5–1.7) for the same pollutant increment. The mechanism by which short-term exposure to SO_2 increases all-cause mortality is unknown.

In an important intervention study from Hong Kong[11] reductions in sulfur content of fuel oil led to a fall in SO_2 concentrations from 44 to 21 $\mu g\,m^{-3}$, particle concentrations were unchanged. Remarkably, the increasing annual death rate (due to ageing of the population) for respiratory diseases fell by 3.9% and for cardiovascular diseases by 2.0%. This has been seen as evidence that long-term exposure to even low concentrations of SO_2 causes cardiorespiratory deaths. Interestingly, the vanadium content of particles also fell – the significance of this is obscure at present.

Time-series studies have also shown that SO_2 is associated with hospital admissions. The World Health Organization report: "*Air Quality Guidelines: A Global Update*"[25] reported a meta-analysis of time-series studies. Table 3 is taken, with permission, from that report.

The consistency of effects on the cardiovascular system is striking and unexpected: SO_2 has been regarded, classically, as a pollutant likely to affect the respiratory system.

Interestingly, a re-analysis of the major US cohort studies showed a clear association between long-term average concentrations of SO_2 and all-cause mortality.[20] All-cause mortality is, in these studies dominated by cardiovascular deaths. Remarkably, the effect of SO_2 remained statistically significant after adjustment for the covariables $PM_{2.5}$ and sulfate (SO_4^{2-}) concentration. Once again, the question of how SO_2 causes such deaths is unanswered: a reflex effect on the control of the heart is at least possible.

Table 3 Summary table based on meta-analysis of time-series. (Reproduced with kind permission of the World Health Organization.)

Pollutant (24 hr average)	N	Outcome measure	Assessment	Random effects (95% CI) (% change per 10 $\mu g/m^3$)
SO$_2$	67	CV mortality	+	0.8 (0.6, 1.0)
SO$_2$	7	CV admissions	+	0.6 (0.1, 1.2)
SO$_2$	18	Cardiac admissions	+	2.4 (1.6, 3.3)
SO$_2$	10	IHD admissions	+	1.2 (0.5, 1.9)
SO$_2$	5	Heart failure admissions		0.9 (−0.1, 1.8)
SO$_2$	7	Cere-brovascular admissions		0.3 (−0.5, 1.1)

7 Ozone

Concentrations of ozone in urban areas of the UK are rising. In part, this is due to an increase in Northern Hemisphere background ozone concentrations and in part to a reduction in ozone scavenging by nitric oxide in urban areas. In the UK, ozone concentrations tend to be higher in rural than in urban areas because of the scavenging effect of nitric oxide; in California, on the other hand, the strong sunlight, high traffic density and production of reactive organic species (see the following equations) leads to ozone concentrations being high in both rural and urban areas. Ozone reacts with nitric oxide to produce nitrogen dioxide:

$$NO + O_3 \rightarrow NO_2 + O_2$$

Downwind of traffic-rich areas, NO$_2$ breaks down under the action of sunlight and this leads to ozone formation:

$$NO_2 + h\nu \rightarrow NO + O^{\cdot}$$
$$O_2 + O^{\cdot} \rightarrow O_3$$
$$RO_2 + NO \rightarrow RO + NO_2$$

Ozone is regarded as an outdoor air pollutant, though in summer, when windows and door are open, indoor levels rise towards those outdoors. Because the formation of ozone is dependent on the formation of traffic-generated precursors and on the action of sunlight, it is not surprising that ozone concentrations reach a peak in the afternoon. The toxicological response to ozone seems to be about equi-dependent on concentration and duration of exposure and peak daily 8-hour average ozone concentrations have been used as a basis (metric) for air quality standards.

The database on the effects of ozone on health is vast.[2] Time-series studies show consistent effects on cardiovascular mortality (See Figure 3). However, no consistent effect on admissions to hospital for treatment of cardiovascular diseases has been found. This lack of coherence in findings is remarkable and unexplained. For respiratory effects the picture is clearer: both daily deaths and admissions to hospital are associated with daily ozone concentrations.

Long-term exposure to ozone has been shown to be related to a slowing of lung development as indicated by poorer than expected performance in lung

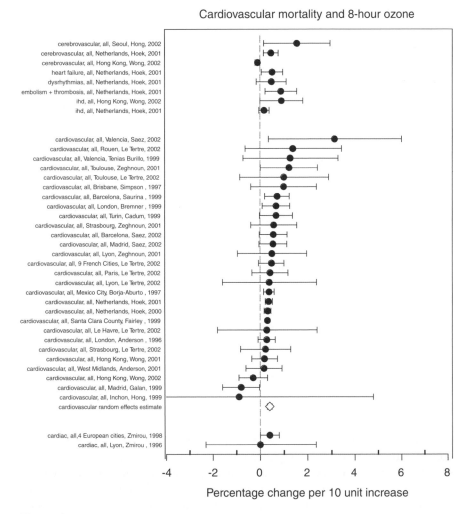

Figure 3 Cardiovascular mortality and 8 hour ozone.

function tests in young people growing up in relatively high ozone areas.[73] Similar results have been reported from the Southern California Children's Health Study.[56,57]

Studies of volunteers exposed to ozone have reported reductions in indices of lung function and an increase in markers of inflammation (cytokines and inflammatory cells) in airway fluid sampled by bronchoalveolar lavage. A detailed account of these studies has been provided.[24,25] Early work suggested a threshold of effect at about 80 ppb ($160 \, \mu g \, m^{-3}$) ozone exposure for 6.6 hours with intermittent exercise. Responses at this level of exposure were small. As with SO_2 and NO_2, the lack of coherence between the results of volunteer studies and those of time-series epidemiological studies is a cause for concern. It may be that the range of sensitivity to these gases is much larger in the general population than in those subjects chosen for volunteer studies.

8 Carbon Monoxide

Carbon monoxide (CO) is, in toxicological terms, the best understood of air pollutants. Carbon monoxide binds, rather remarkably, to haemoglobin (Hb) in precisely the same way as oxygen – the CO–Hb dissociation curve is identical in shape with the O_2–Hb dissociation curve. But the affinity of CO for haemoglobin is > 200 greater than that of oxygen and thus low concentrations of CO compete effectively with ambient concentrations of oxygen for binding to haemoglobin. Oxygen transport is impaired and the O_2–Hb dissociation curve is shifted making release of O_2 at the tissues less effective.

The reason for displacement of the dissociation curve is easy to grasp. Normally, fully oxygenated Hb contains 4 molecules of oxygen and as each is released, as the P_{O2} falls, the "next to be released" is released more easily due to "cooperation" – a homely term to explain complex changes in conformation. In CO poisoning, many molecules of Hb carry only one or two molecules of O_2 and thus the capacity for "cooperation" is reduced. In fact, the oxygen dissociation curve for Hb carrying a mixture of 2 CO molecules and 2 O_2 molecules looks like the upper part of the normal curve and the steep portion is "missing". This so-called left shift of the dissociation curve is particularly important in the fetus: fetal Hb has an already left-shifted oxygen dissociation curve and further displacement to the left seriously impairs oxygen release.

Exposure to CO that causes 2.3% of haemoglobin to occur as COHb (COHb is measured as a percentage of total Hb) impairs oxygen delivery to the myocardium and patients suffering from poor coronary circulation experience chest pain earlier than expected during exercise. Studies in which volunteers suffering from angina due to impaired blood supply to the heart were exposed to low concentrations of carbon monoxide, and asked to exercise on a treadmill, have shown that chest pain occurred earlier during exercise than when ambient air was breathed. Furthermore, the characteristic signs of a reduction in oxygen supply to the heart seen on an electrocardiogram during exercise in such subjects occurred earlier than expected.[23]

Despite this mechanism, whether exposure to ambient concentrations of CO damages health is unknown and regarded by many as toxicologically unlikely: concentrations of CO are generally too low to produce significant changes in COHb levels. Epidemiological studies have shown associations between daily variations in CO concentrations and effects on the cardiovascular system[9] but the close correlation between ambient concentrations of CO and fine particles makes interpretation of these findings difficult. Long-term exposure to CO, *e.g.* in cigarette smokers has been found to be associated with atheromatous changes in the coronary arteries, but there is little evidence that such an effect occurs at ambient concentrations.[2]

9 Carcinogenic Air Pollutants found in Urban Areas

Urban air contains a number of carcinogenic air pollutants. These include benzene, 1,3-butadiene and the polycyclic aromatic hydrocarbon compounds (PAH). In addition, small amounts of inorganic carcinogens such as arsenic (produced by coal burning) are present. Details of the formation of the organic carcinogens are provided elsewhere but it is worth noting here that these compounds range in molecular weight and, at ambient temperature, from gases to solids. Some, including the higher molecular weight PAH compounds are condensed onto the surface of particles and may account for the carcinogenicity of the urban aerosol previously reported.[17] It is generally accepted that no threshold of effect can be identified regarding genotoxic carcinogens: it is assumed that exposure to low concentrations cause a small increase in the risk of cancer. The fundamental mechanisms of action of genotoxic carcinogens support the view that no threshold of effect can be identified for these compounds.

Given that no threshold of effect can be identified, it is clearly impossible to recommend an air quality guideline or set an air quality standard that guarantees absolute safety. Standards have, however, been recommended in the UK;[74–77] the thinking behind the approach used has since been explained.[78] In essence, the approach depends on identifying a lowest observed adverse effect level (LOAEL) or a no observed adverse effect level (NOAEL) from the literature. If only a LOAEL could be identified this was divided by 10 to produce an estimated NOAEL. This value was then converted to an estimated lifetime exposure NOAEL:

If the NOAEL were to be expressed in terms of an occupational exposure (8-hour per day, 5 days per week) as x μg m^{-3} for y years, then the estimated lifetime exposure NOAEL would be:

$$x \times \frac{y}{70} \times \frac{8}{24} \times \frac{240}{365}$$
$$\textit{i.e. } 0.00313 \, xy$$

If a working lifetime (40 years) equivalent exposure is calculated then this can be divided by 10 to produce an approximate estimated whole life equivalent exposure.[74] A further factor of 10 was then applied to allow for the possible range of sensitivity in the general population. For benzene this led to a proposed standard of 5 ppb ($15.95\,\mu g\,m^{-3}$) expressed as a running annual average concentration.

The approach described is similar to that used by the World Health Organization in calculating Unit Risk Factors: *i.e.*, the additional risk associated with lifetime exposure to unit concentration of genotoxic carcinogens (in the case of benzene, $1\,\mu g\,m^{-3}$). This is expressed by the equation:

$$UR = \frac{PoR - Po}{X}$$

where:

P_o is the background lifetime risk of the effect (disease) in question in the specified population;

R is the Relative risk in the reference group (the group exposed occupationally to the carcinogen): *i.e.* R=Observed number of cases/Expected number of cases;

X is the correction factor need when considering lifetime exposure as compared with the exposure leading to Relative Risk, R.

$$X = x \times \frac{8}{24} \times \frac{240}{365} \times \frac{y}{70}$$

In the World Health Organization Air Quality Guidelines[25] the Unit Risk for benzene was given as, for a lifetime exposure to $1\,\mu g\,m^{-3}$, 6×10^{-6}. Both approaches depend on good quality studies of the effects of occupational exposure to the compound concerned. Effects cannot be easily detected at environmental levels: very large and expensive studies would be needed. The two approaches described above are similar in that they ignore the possible alinearity of the exposure response curve for carcinogens: it is assumed that long-term exposure to a low concentration is as dangerous as short-term exposure to a proportionately higher concentration. More sophisticated methods are available, though proof of their greater reliability is, in general, lacking. The UK approach suffers from encouraging the reader to think that attainment of the recommended standard confers complete safety: that this is not the correct interpretation was explained at some length in the EPAQS reports.

References

1. Dept. Environ., Food and Rural Affairs, *An Economic Analysis to Inform the Air Quality Strategy. Updated Third Report of the Interdepartmental Group on Costs and Benefits*, Defra, London, 2007, vol. 1–3.
2. World Health Organization, *Global Burden of Disease and Risk Factors*, World Band and Oxford University Press, New York, 2006.

3. J. Spengler, J. M. Samet and J. F. McCarthy, *Indoor Air Quality Handbook,* McGraw-Hill, New York, NY, 2001.
4. H. Özkaynak and J. Spengler, in *Particles in Our Air,* ed. R. Wilson, R. and J. Spengler, Harvard University Press, Cambridge, MA, 1996, pp. 63–84.
5. K. Steenland and D. Savitz, *Topics in Environmental Epidemiology,* Oxford University Press, New York, NY, 1997.
6. F. Dominici, S. L. Zeger and J. M. Samet, *Res. Rep. Health. Eff. Inst.*, 2000, **94**.
7. J. Heinrich, U. Gehring, J. Cyrys, M. Brauer, G. Hoek, P. Fischer, T. Bellander and B. Brunekreef, *Occup. Environ. Med.*, 2005, **62**, 517–523.
8. R. E. Waller, in *Oxford Textbook of Public Health,* 2nd edn. W. W. Holland, R. Detels and G. Knox, Oxford University Press, 1991, **vol. 2**, pp. 435–450.
9. Dept. Health, Committee Med. Effects Air Pollutants, *Cardiovascular Disease and Air Pollution*, Dept. Health, London, 2006.
10. L. Clancy, P. Goodman, H. Sinclair and D. W. Dockery, *Lancet*, 2002, **360**, 1210–1214.
11. A. J. Hedley, C. M. Wong, T. Q. Thach, S. Ma, T. H. Lam and H. R. Anderson, *Lancet*, 2002, **360**, 1646–1652.
12. M. S. Friedman, K. E. Powell, L. Hutwagner, L. M. Graham and W. G. Teaque, *JAMA*, 2001, **285**, 897–905.
13. C. A. Pope, *Am. J. Public Health*, 1989, **79**, 623–628.
14. C. A. Pope, *Arch. Environ. Health*, 1991, **46**, 90–97.
15. D. W. Dockery, C. A. Pope, X. Xu, J. D. Spengler, J. H. Ware, M. E. Fay, B. G. Ferris and F. E. Speizer, *N. Engl. J. Med.*, 1993, **329**, 1753–1759.
16. C. A. Pope, M. J. Thun, M. M. Namboodiri, D. W. Dockery, J. S. Evans, F. E. Speizer and C. W. Heath, *Am. J. Respir. Crit. Care Med.*, 1995, **151**, 669–674.
17. C. A. Pope, R. T. Burnett, M. J. Thun, E. E. Calle, D. Krewski, K. Ito. and G. D. Thurston, *JAMA*, 2002, **287**, 1132–1141.
18. C. A. Pope, R. T. Burnett, G. D. Thurston, M. J. Thun, E. E. Calle, D. Krewski and J. J. Godleski, *Circulation*, 2004, **109**, 71–77.
19. Dept. Health, Committee Med. Effects Air Pollutants, *Long-Term Exposure to Air Pollution: Effect on Mortality*, Dept. Health, London, 2007.
20. D. Krewski, R. T. Burnett, M. S. Goldberg, K. Hoover, J. Siemiatycki, M. Jerrett, M. Abrahamowicz and W. H. White, *Reanalysis of the Harvard Six Cities Study and the American Cancer Society Study of Particulate Air Pollution and Morbidity,* Health Effects Institute, Boston, MA, 2000.
21. M. Jerrett, R. T. Burnett, R. Ma, C. A. Pope, D. Krewski, K. B. Newbold, G. Thurston, Y. Shi, N. Finkelstein, E. E. Calle and M. J. Thun, *Epidemiology*, 2005, **16**, 727–736.
22. G. Hoek, B. Brunekreef, S. Goldbohm, P. Fischer and P. A. Van den Brandt, *Lancet*, 2002, **360**, 1203–1209.
23. S. T. Holgate, J. M. Samet, H. S. Koren and R. L. Maynard, *Air Pollution and its Effects on Health,* Academic Press, London, New York, 1999.
24. M. Lippmann, *Environmental Toxicants: Human Exposures and their Health Effects,* Wiley-Interscience, New York, 2000.

25. World Health Organization, *Air Quality Guidelines. Global Update 2005. Particulate Matter, Ozone, Nitrogen Dioxide and Sulfur Dioxide*, World Health Organization, Copenhagen, 2006.
26. G. Oberdörster, E. Oberdörster and J. Oberdörster, *Environ. Health Perspect.*, 2005, **113**, 823–839.
27. M. Lippmann, T. Gordon and L. C. Chen, *Inhalation Toxicol.*, 2005, **17**, 199–207.
28. M. Lippmann, T. Gordon and L. C. Chen, *Inhalation Toxicol.*, 2005, **17**, 255–261.
29. A. Seaton, W. MacNee, K. Donaldson and D. Godden, *Lancet*, 1995, **345**, 176–178.
30. G. Oberdörster, R. M. Gelein, J. Ferin and B. Weiss, *Inhalation Toxicol.*, 1995, **7**, 111–124.
31. J. R. Batalha, P. H. Saldiva, R. W. Clarke, B. A. Coull, R. C. Sterns, J. Lawrence, G. G. Murthy, P. Koutrakis and J. J. Godleski, *Environ. Health Perspect.*, 2002, **110**, 1191–1197.
32. J. J. Godleski, R. L. Verrier, P. Koutrakis, P. Catalano, B. Coull, U. Reinisch, E. G. Lovett, J. Lawrence, G. G. Murthy, J. M. Wolfson, R. W. Clarke, B. D. Nearing and C. Killingsworth, *Res. Rep Health Eff. Inst.*, 2000, **91**, 5–88.
33. A. Nemmar, H. Vanbilloen, M. F. Hoylaerts, P. H. M. Hoet, A. Verbruggen and B. Nemery, *Am. J. Respir. Crit. Care Med.*, 2001, **164**, 1665–1668.
34. A. Nemmar, P. H. M. Hoet, B. Vanquickenborne, D. Dinsdale, M. Thomeer, M. F. Hoylaerts, H. Vanbilloen, L. Mortelmans and B. Nemery, *Circulation*, 2002, **105**, 411–414.
35. A. Nemmar, M. F. Hoylaerts, P. H. M. Hoet, D. Dinsdale, T. Smith, H. Xu, J. Vermylen and B. Nemery, *Am. J. Respir. Crit. Care Med.*, 2002, **166**, 998–1004.
36. A. Nemmar, M. F. Hoylaerts, P. H. M. Hoet, J. Vermylen and B. Nemery, *Toxicol. Appl. Pharmacol.*, 2003, **186**, 38–45.
37. A. Nemmar, P. H. M. Hoet, D. Dinsdale, J. Vermylen, M. F. Hoylaerts and B. Nemery, *Circulation*, 2003, **107**, 1202–1208.
38. T. Suwa, J. C. Hogg, K. B. Quinlan, A. Ohgami, R. Vincent and S. F. Van Eeden, *J. Am. Coll. Cardiol.*, 2002, **39**, 935–942.
39. L. C. Chen and J. S. Hwang, *Inhalation Toxicol.*, 2005, **17**, 209–216.
40. L. C. Chen and C. Nadziejko, *Inhalation Toxicol.*, 2005, **17**, 217–224.
41. J. S. Hwang, C. Nadziejko and L. C. Chen, *Inhalation Toxicol.*, 2005, **17**, 199–207.
42. A. Gunnison and L. C. Chen, *Inhalation Toxicol.*, 2005, **17**, 225–233.
43. P. Maciejczyk and L. C. Chen, *Inhalation Toxicol.*, 2005, **17**, 243–253.
44. P. Maciejczyk, M. Zhong, Q. Li, J. Xiong, C. Nadziejko and L. C. Chen, *Inhalation Toxicol.*, 2005, **17**, 189–197.
45. B. Veronesi, O. Makwana, M. Pooler and L. C. Chen, *Inhalation Toxicol.*, 2005, **17**, 235–241.
46. R. B. Devlin, A. J. Ghio, H. Kehrl, G. Sanders and W. Cascio, *Eur. Respir. J.*, 2003, **40**(Suppl), 76s–80s.

47. A. Peters, S. Perz, A. Doring, J. Stieber, W. Koenig and H. E. Wichmann, *Am. J. Epidemiol.*, 1999, **150**, 1094–1098.
48. A. Peters, E. Liu, R. L. Verrier, J. Schwartz, D. R. Gold, M. Mittleman, J. Baliff, J. A. Oh, G. Allen, K. Monahan and D. W. Dockery, *Epidemiology*, 2000, **11**, 11–17.
49. D. M. Stieb, S. Judek and R. T. Burnett, *J. Air Waste Manage. Assoc.*, 2002, **52**, 470–484.
50. J. M. Samet, S. L. Zeger, F. Dominici, F. Curriero, I. Coursac, D. W. Dockery, J. Schwartz and A. Zanobetti, *Res. Rep. Health Eff. Inst.*, 2000, **94**(Pt 2), 5–70; discussion 71–79.
51. F. Dominici, A. McDermott, M. Daniels, S. L. Zeger and J. M. Samet, in *Revised Analyses of Time-Series Studies of Air Pollution and Health (Health Effects Institute Special Report)*, Capital City Press, Montpelier, VT, 2003, 9–24.
52. K. Katsouyanni, J. Schwartz, C. Spix, G. Touloumi, D. Zmirou, A. Zanobetti, B. Wojtyniak, J. M. Vonk, A. Tobias, A. Ponka, S. Medina, L. Bacharova and H. R. Anderson, *J. Epidemiol. Community Health*, 1996, **50**, S12–S18.
53. H. R. Anderson, C. Spix, S. Medina, J. P. Schouten, J. Castellsague, G. Rossi, D. Zmirou, G. Touloumi, B. Wojtyniak, A. Ponka, L. Bacharova, J. Schwartz and K. Katsouyanni, *Thorax*, 1997, **52**, 760–765.
54. C. Spix, H. R. Anderson, J. Schwartz, M. A. Vigotti, A. LeTertre, J. M. Vonk, G. Touloumi, F. Balducci, T. Piekarski, L. Bacharova, A. Tobias, A. Ponka and K. Katsouyanni, *Arch. Environ. Health*, 1998, **53**, 54–56.
55. J. L. Peel, P. E. Tolbert, M. Klein, K. B. Metzger, W. D. Flanders, K. Todd, J. A. Mulholland, P. B. Ryan and H. Frumkin, *Epidemiology*, 2005, **16**, 164–174.
56. W. J. Gauderman, R. McConnell, F. Gilliland, S. London, D. Thomas, E. Avol, H. Vora, K. Berhane, E. B. Rappaport, F. Lurmann, H. G. Margolis and J. Peters, *Am. J. Respir. Crit. Care Med.*, 2000, **162**, 1383–1390.
57. W. J. Gauderman, G. F. Gilliland, H. Vora, E. Avol, D. Stram, R. McConnell, D. Thomas, F. Lurmann, H. G. Margolis, E. B. Rappaport, K. Berhane and J. M. Peters, *Am. J. Respir. Crit. Care Med.*, 2002, **166**, 76–84.
58. R. McConnell, K. Berhane, F. Gilliland, J. Molitor, D. Thomas, F. Lurmann, E. Avol, W. J. Gauderman and J. M. Peters, *Am. J. Respir. Crit. Care Med.*, 2003, **168**, 790–797.
59. C. Schindler, U. Ackerman-Liebrich, P. Leuenberger, C. Monn, R. Rapp, G. Bolognini, J. P. Bongard, O. Brändii, G. Domenighetti, W. Karrer, R. Keller, T. G. Medici, A. P. Perruchoud, M. H. Schoni, J. M. Tschopp, B. Villiger and J. P. Zellweger, *Epidemiology*, 1998, **9**, 405–411.
60. G. L. Snider, J. Kleinerman, W. M. Thurlbeck and Z. H. Bengali, *Am. Rev. Respir. Dis.*, 1985, **32**, 182–185.
61. E. L. Avol, W. S. Linn, R. C. Peng, J. D. Whynot, D. A. Shamoo, D. E. Little, M. N. Smith and J. D. Hackney, *Toxicol. Ind. Health*, 1989, **5**, 1025–1034.

62. M. A. Bauer, M. J. Utell, P. E. Morrow, D. M. Speers and F. R. Gibb, *Am. Rev. Respir. Dis.*, 1986, **134**, 1203–1208.
63. L. J. Roger, D. H. Horstman, W. McDonnell, H. Kerhl, P. J. Ives, E. Seal, R. Chapman and E. Massaro, *Toxicol. Ind. Health*, 1990, **6**, 155–171.
64. R. Jörres, D. Nowak, F. Grimminger, W. Seeger, M. Oldigs and H. Magnussen, *Eur. Respir. J.*, 1995, **7**, 1213–1220.
65. T. Sandström, R. Helleday, L. Brermer and N. Stjernberg, *Eur. Respir. J.*, 1992, **5**, 1092–1098.
66. H. A. Boushey, I. Rubinstein and B. G. Bigby, *Studies on Air Pollution: Effects of Nitrogen Dioxide on Airway Caliber and Reactivity in Asthmatic Subjects; Effects of Nitrogen Dioxide on Lung Lymphocytes and Macrophage Products in Healthy Subjects; Nasal and Bronchial Effects of Sulfur Dioxide in Asthmatic Subjects.* (Report No ARB/R-89/384), California Air Resources Board, Sacramento, CA, 1988.
67. M. O. Amdur, in *Casarett and Doull's Toxicology: The Basic Science of Poisons,* ed. C. D. Klaassen, M. O. Amdur and J. Doull, Macmillan, London, New York, Toronto, 3rd edn. 1986, pp. 801–824.
68. W. S. Linn, T. G. Venet, D. A. Shamoo, L. M. Valencia, U. T. Anzar, C. E. Spier and J. D. Hackney, *Am. Rev. Respir. Dis.*, 1983, **127**, 278–283.
69. W. S. Tunnicliffe, M. F. Hilton, R. M. Harrison and J. G. Ayres, *Eur. Respir. J.*, 2001, **17**, 604–608.
70. K. Katsouyanni, G. Touloumi, C. Spix, J. Schwartz, F. Balducci, S. Medina, G. Rossi, B. Wojtyniak, J. Sunyer, L. Bacharova, J. P. Schouten, A. Ponka and H. R. Anderson, *BMJ*, 1997, **314**, 1658–1663.
71. E. Samoli, J. Schwartz, B. Wojtyniak, G. Touloumi, C. Spix, F. Balducci, S. Medina, G. Rossi, J. Sunyer, L. Bacharova, H. R. Anderson and K. Katsouyanni, *Environ. Health Perspect.*, 2001, **109**, 349–353.
72. F. Ballester, M. Saez, S. Perez-Hoyos, C. Iniguex, A. Gandarillas, A. Tobias, J. Bellido, M. Taracido, F. Arribas, A. Daponte, E. Alonso, A. Canada, F. Guillen-Grima, L. Cirera, M. J. Perez-Boillos, C. Saurina, F. Gomez and J. M. Tenias, *Occup. Environ. Med.*, 2002, **59**, 300–308.
73. A. Galizia and P. Kinney, *Environ. Health Perspect.*, 1999, **107**, 675–679.
74. Dept. Environ, *Expert Panel on Air Quality Standards. Benzene.* HMSO, London, 1994.
75. Dept. Environ. *Expert Panel on Air Quality Standards. 1,3-Butadiene,* (*First Report*), HMSO, London, 1994.
76. Dept. Environ, Transport and The Regions, *Expert Panel on Air Quality Standards. Polycyclic Aromatic Hydrocarbons*, The Stationery Office, London, 1999.
77. Dept. Environ, Food and Rural Affairs, *Expert Panel on Air Quality Standards. Second Report on 1,3-Butadiene,* Defra Publications, London, 2002.
78. R. L. Maynard, K. M. Cameron, R. Fielder, A. McDonald and A. Wadge, *Regulat. Toxicol. Pharmacol.*, 1997, **26**, S60–S70.

The Policy Response to Improving Urban Air Quality

MARTIN WILLIAMS

ABSTRACT

This chapter provides a brief historical perspective on the policy measures and their effect on improving urban air quality over the past few decades. The emphasis has shifted from fairly crude source related controls, involving curbs on the use of coal in the formative years of urban air quality management in the 1956 Clean Air Act, through to the combination of source related emission standards for vehicles and larger industrial emitters in combination with a series of air quality standards which we currently use. As our knowledge of the atmospheric science of air pollutants has improved and as our knowledge of their harmful effects has become increasingly quantitative, more sophisticated urban air quality management systems have become practicable. A 'third generation' of air quality management has been developed using advances in epidemiology in combination with more widespread air quality monitoring together with modelling, to develop an approach known as 'exposure reduction' to target air quality management to deliver more effective means of reducing the adverse impacts on public health on a wider scale than simply using a single air quality standard. The chapter finally discusses the challenge of maximising the co-benefits of climate change and air quality policies, and minimising the adverse effects of one set of policies on the other.

Issues in Environmental Science and Technology, 28
Air Quality in Urban Environments
Edited by R.E. Hester and R.M. Harrison
© Royal Society of Chemistry 2009
Published by the Royal Society of Chemistry, www.rsc.org

1   Introduction

Urban air quality in the UK has improved significantly in the past 50 years. The policy responses have also changed profoundly during this period in many respects, and indeed are still developing, but over the past decades they have changed in one fundamentally important way. Prior to, and including, the 1956 Clean Air Act, pollution control in the UK had been concerned since the late 19th century with *emissions* at source and was predominantly focussed on industrial sources. Even though the 1952 Smog arose from widespread diffuse sources, there was no mention in the 1956 Act of air concentrations, standards or guidelines, nor was there any explicit recognition in policy instruments of any concept of exposure to people or the environment. Contrast this with the present day where the emphasis of policy has shifted almost completely to air *quality* targets of one form or another, with legislation on emissions forming the means of *implementation* and the achievement of the agreed environmental quality. Along the way there have been many reasons for this change and many associated developments in policy. Not least among these have been the rise in importance of the European Union and other international fora, and the much wider availability of information and data and involvement of a much larger constituency of 'stakeholders' on both sides of the environmental debate, in business and in the environmental organisations. This paper, which cannot be comprehensive in the space available, seeks to draw out the main events in the evolution of policy responses to improving urban air quality in the UK. The discussion will not simply concern itself with the policy instruments themselves but will also address the significant steps taken to formulate the evidence (from atmospheric science, epidemiology, toxicology and from economics) which has informed and underpinned the policy developments over the past fifty years or so.

2   Policy and the Evidence Base after the Smog Episodes of 1952 and 1962

The Clean Air Act of 1956 had several important elements. Firstly it embodied a considerable element of devolution of action to local authorities, who were empowered to designate Smoke Control Areas which required the use of either authorised solid fuels or authorised appliances in the domestic sector. The authorisation was, and still is, based on emission limits set, at the national level, such as to effectively prohibit the emission of smoke at the levels of uncontrolled burning before the Act. Grants were payable to ease the transition away from open fires. The major cities of the UK became virtually completely smoke-controlled and emissions of smoke and sulfur dioxide decreased quite significantly. The process took time however, progress was not fast enough to prevent the smog episode of December 1962 from occurring when weather conditions similar to those in the 1952 episode prevailed, even though domestic coal consumption in London was half what it had been in 1952. Emissions decreased very quickly after that however, and by 1991, when similar meteorological conditions occurred in London in December, mean black smoke levels in London had dropped by almost 90% of the 1962 levels. The effects of the 1991 air pollution episode were much less than those of the smog of 1952. In 1952, mortality rates in London, calculated over a two week period, rose by 230%; the increase reported for 1991 was 10%.

It is fair to say that on their own the provisions of the Clean Air Act would not have lead to as large a decrease in urban emissions of sulfur dioxide as occurred. The main solid fuel replacement for coal, so-called smokeless fuel, had a sulfur content not much lower than UK coal, even though most of the volatiles which formed black smoke on combustion had been removed. The other factor which occurred at around the same time was the general move away from the use of solid fuel for heating to the use of gas, or even electricity to heat premises. Both resulted in significant reductions in smoke and sulfur dioxide concentrations in urban areas, although in the case of electric heating the emissions were generated in power stations most of which had been moved from city centres to rural areas. These large extra-urban power plants, with tall stacks helped to reduce urban concentrations considerably, although before abatement technology was installed to reduce emissions they contributed to problems via 'acid rain' in more remote areas, not just in Scandinavia, but in areas of the UK too.

During this period, there was a large amount of activity on the scientific front. The National Survey of Air Pollution was set up in the early 1960s (with data still available on the Defra archive www.airquality.co.uk from around 1962), with measurements of black smoke and sulfur dioxide being made by local authorities. Co-ordination, quality control and data summaries and analysis were carried out by the newly established Warren Spring Laboratory. This network used the relatively cheap and simple smoke stain/bubbler method (which strictly speaking measured total acidity rather than being specific for sulfur dioxide) and in the first two decades of its existence was composed of over 1,000 monitoring sites. Around this time, the Medical Research Council established the Air Pollution Unit to undertake research into the effects of air pollutants on health under the leadership of Professor Pat Lawther, who together with Robert Waller and co-workers established the first epidemiological and toxicological studies linking health effects with air pollution concentrations. This science base of air pollution concentrations covering the whole of the UK, and knowledge of their effects on health represented the most important information resource on the effects of smoke and sulfur dioxide in Europe and probably the world.

This however proved to be a two-edged sword; as the 1970s drew to a close, the control of coal emissions was clearly being successful and this was considered to have largely solved the problem of the health effects of urban air pollution. The world renowned Air Pollution Unit of the Medical Research Council was closed in 1978. In the USA, the US Clean Air Act was promulgated in 1970, and this set standards for other pollutants, mostly related to motor vehicle emissions, such as carbon monoxide, lead, oxides of nitrogen and ozone, as well as for sulfur dioxide and particles. In the mid 1970s policy makers in the UK also began to recognise the growing importance of pollution from motor vehicles and monitoring stations were set up to measure traffic related pollutants in a small number of locations. These used the newly emerging automatic monitoring techniques to record hourly data and also addressed lead, carbon monoxide, oxides of nitrogen and ozone. At this time, however, there was no further policy activity; no health effects studies were being carried out on vehicle related pollutants and there was no particular incentive to produce air quality management legislation to cover the 'newer' pollutants.

Indeed, attention was diverted away from urban air pollution in the late 1970s and early 1980s by the recognition of the problems of long range transport of air pollution, acid rain and effects on more rural and remote ecosystems. A considerable amount of policy-making resource was devoted to solving these problems. This is not the place to discuss the detailed history of that problem, but suffice it to say international scale issues of long-range transboundary air pollution (LRTAP) dominated policy for most of the 1980s. The early Scandinavian work in the OECD (Organisation for Economic Co-operation and Development) was taken up by the UN Economic Commission for Europe (UNECE) leading to the signing of the Convention on LRTAP in 1979. There followed a series of Protocols initially addressing the acid rain issue but subsequently dealing with ozone and its precursors and more recently heavy metals and persistent organic pollutants.

One outcome of this process with benefits for urban air quality was the fact that this development was instrumental in opening up information on emissions and air quality across Europe, including the UK. To today's audience, where even the most detailed data are freely available on the Internet, this may not sound a major achievement, but at the time of the negotiation of the Convention, in some countries even national emission totals were considered to be extremely sensitive and were treated almost as state secrets.

Notwithstanding the acid rain issue, during the late 1970s and early '80s, the policy process in the European Community was beginning to get into its stride, and began to formulate legislation based on health related problems of air pollution. This has represented a major development in European air quality policy and probably represents one of the most important changes in policy making in Europe in the last fifty years. From this point onwards, the urban air quality policy making process in Europe has been dominated by the production of legislation within the EU, and to some extent the UNECE. Moreover there has also been a considerable amount of co-operative activity to develop the atmospheric science base, the health effects and the economics underpinning these policy developments. It has been a long and often difficult process, and continues to be a challenge, as the Member States have had to adapt to the principles of collective decision making in the EU and in other international fora.

During the 1980s a major initiative took place within the European Regional Office of the World Health Organisation, culminating in 1987 with the publication of the WHO Air Quality Guidelines for Europe.[1] This addressed the effects on human health from 28 air pollutants and recommended guideline concentrations for averaging times of relevance to the effect in question. From a policy perspective it is very important to be clear about the nature of these guidelines. It would first help to distinguish between risk assessment and risk management. In dealing with air quality problems, and many others, the process can be simplified for clarity to one where the risk from a particular pollutant is assessed. For an air pollutant, if sufficient health effect and air concentration information is available, this might take the form of an assessment of the total burden of a given health outcome over a country. This result is determined purely by science and does not take into account any consideration of how much pollution could or should be reduced. The management of this risk is the phase where these issues are decided and other factors come into play such as the technical feasibility of reducing emissions and exposures, the associated costs, the monetisation of the health effects themselves where this is possible (and in some regions this could even be an ethical or moral issue). The policy maker, and ultimately society as a whole, has to balance all these factors in managing the risk in terms of proportionate actions. Depending on the balance of the stringency of control measures, or costs, it is possible that the risk may not be completely eliminated, and society *de facto* accepts some residual adverse health effects. The WHO Guidelines form part of the risk *assessment* process. They were originally formulated to indicate concentrations and averaging times which would not lead to adverse health effects from the pollutant concerned. They are thus different in nature from legally enforceable air quality standards or limit values, the agreement on which has to balance the costs and the risks in formulating acceptable policy solutions. The WHO Guidelines however have been extremely influential in the European policy process as they have generally formed the starting point for negotiations on limit values for air pollutants in European legislation.

With the commencement of activity on the part of the European Commission, the first air quality Directive was agreed in 1980 and dealt with particulate matter ('smoke') and sulfur dioxide. The limit values were based on the health effect work done in the UK at the MRC Unit mentioned previously. This Directive was followed by others dealing with Lead and Nitrogen Dioxide. The evidence base for adverse effects from these pollutants was mixed. There was a considerable amount of epidemiological and toxicological work done on lead, and the issue generated a large amount of controversy. There was little need for very subtle science in analysing the causes and effects of the 1952 smog, but the lead debate in the 1970s was the first time that relatively sophisticated medical science and statistics had played a role in influencing public attitudes to air pollution problems and subsequently in influencing policy.

The Directive on nitrogen dioxide agreed in 1985, represented a milestone in air quality policy in the EU as it was the first to recognise explicitly (even though the text was buried in an Annex) the concept that Limit Values should apply only in locations where people are likely to be exposed over time periods for which adverse effects might occur. At the time of these Directives there was a degree of debate on this issue and the thinking behind the insertion of this 'exposure' criterion was to prevent measurements being taken and breaches of the limit values alleged, at locations which would be legally within the scope of the Directive if no exposure criterion were invoked but which were clearly nonsensical in terms of public exposure, like the central area between two carriageways of a motorway or major road.

The thinking behind this approach was also reflected in the UK's support for effects-based instruments to address the acid rain issue. The UK did not sign the first UNECE Protocol, which set what was felt to be an arbitrary reduction of 30% in sulfur emissions. In the event the UK achieved the reduction, and more, but was active in promoting an approach which sought to set targets more closely related to environmental effects through the 'critical loads' concept. This resulted in the major achievement of the Second Sulfur Protocol in 1994, which set emission targets for countries which were related to their contribution to environmental problems, and has formed the basis for subsequent Protocols addressing the same issues.

3   The Development of Strategies in the UK and Europe

While most of the focus of operational air quality policy in the 1980s was on acid rain, and those EC Directives in force dealing with smoke, sulfur dioxide, lead and nitrogen dioxide, there was also a great deal of strategic thinking under way which lead in September 1990 to the publication of a comprehensive white paper setting out the UK's strategy for the environment. 'This Common Inheritance'[2] not only broke with tradition for Government White Papers by the use of colour photographs, charts and maps, it also painted a very broad canvas, and set the UK's stance on environmental issues in a global context for the first time. On air quality, it recognised the growing concern once more over the health effects of air pollutants, and announced the establishment of an expert panel to set standards for the UK. This became the Expert Panel on Air Quality Standards (EPAQS), which has played such an important role in UK air quality policy in the past decade. It also stated explicitly that the public should have a right of access to information held by pollution control authorities, and stressed the role of monitoring as a fundamental activity underpinning the development and evaluation of policy. Alongside this more strategic approach to air quality and environmental issues generally, the Environmental Protection Act of 1990 (EPA 1990) included an updating of previous regulation of industrial emissions, discussed further below. However, it was during the period running up to the EPA 1990 that major changes were taking place in the UK's industrial sectors with a general reduction in the older 'smokestack' manufacturing industries, including iron and steel and related activities. This resulted in major reductions in emissions from this sector and while much of this industry was outside major urban areas, the changes nonetheless contributed to improvements in urban air quality, particularly as regards sulfur dioxide and particulate emissions.

One of the main motivating forces behind the strategic thinking on urban air quality at this time is the fact that it was during this period, in the late 1980s and early 1990s, that major advances in the elucidation of the adverse health effects of urban air pollution were being made. These advances arose chiefly from the epidemiological studies of Schwartz, Dockery and co-workers in the USA, who showed, in a series of influential papers, that levels of particulate matter, which would previously have been considered harmless, were in fact associated not only with ill-health but with mortality. These studies employed sensitive and sophisticated statistical techniques and initially were greeted with scepticism in some quarters. However, further studies by other groups and intensive reanalysis and scrutiny by independent researchers, notably the Health Effects Institute in the USA, have led to these studies not only being accepted, but their findings used to inform policy development in the USA, in the EU and elsewhere. These issues engendered a growing interest in urban air quality and health effects, after ten to fifteen years when few advances had been made.

Accountability for delivering on policies was an issue in the years immediately following 'This Common Inheritance', and several annual follow-up reports set out the Government's performance against the commitments in the original white paper. The context of air pollution policy, and that of environmental policy more generally was broadening following the Brundtland Report in 1987, and in 1994 the Government published its first Sustainable Development Strategy, which was followed by a further version in 1999.[3,4] This went further than the 1990 White Paper and, *inter alia*, identified urban air quality as a key issue.

In parallel with this development, a more strategic approach was being taken to urban air quality issues. Two key publications, in 1994 and 1995, respectively, were 'Improving Air Quality'[5] and 'Air Quality: Meeting the Challenge'[6] which launched proposals for air quality standards and management systems, including the concept of Local Air Quality Management, which in many ways developed the concept of local approaches to air quality management inherent in the smoke control legislation of the 1956 Clean Air Act. These documents began the process which led to the Environment Act of 1995, and which in turn produced, in March 1997, the publication of the UK Air Quality Strategy (AQS),[7] required by the Act.

The 1995 Environment Act was important in several ways. Space does not permit an adequate appraisal of the immense amount of scientific and medical research and assessments which have underpinned the AQS and successors since the 1980s. Aside from the research programmes of the Departments of Environment (in its various guises) and Health, and the research councils, particularly the Natural Environment Research Council, there have been a series of definitive synthesis reports from expert groups in both departments. Reports produced by the Photochemical Oxidants Review Group, the Quality of Urban Air RG, the Airborne Particles RG and the series of reports from the DH Committee on the Medical Effects of Air Pollutants (COMEAP) and its forerunner the group on the Medical Aspects of Air Pollution Episodes

(MAAPE), have been immensely influential in the wider European and inter-national arenas as well as in the UK, and have been highly regarded by both scientists and policy makers. The reports of the joint DoE/DH group EPAQS in particular, have helped establish the UK as a leading force in air quality policy development, and have played an important role in informing the development of EU policy as well as that in the UK.

Two of the most important reports of the COMEAP group were the report on the health effects of particulate matter published in 1995.[8] It has already been noted above that the initial time-series studies of Schwartz, Dockery and others in the USA were treated with some scepticism at first. The COMEAP report provided an early objective appraisal of the growing body of epide-miological evidence of the potential harmful effects of exposure to particulate matter, and concluded that 'The Committee considers that the reported asso-ciations between daily concentrations of particles and acute effects on health principally reflect a real relationship and not some artefact of technique or the effect of some confounding factor.

In terms of protecting public health it would be imprudent not to regard the demonstrated associations between 'daily concentrations of particles and acute effects on health as causal'. This advice provided an extremely useful steer to the policy making process, and coupled with the advice from the EPAQS panel in recommending air quality standards (in the sense of the WHO guidelines, as levels below which one might not expect to see harmful effects in the population as a whole) and the production of the major Air Quality Strategy, provided a degree of leadership in Europe which was subsequently built upon by the European Commission as discussed below. The other significant report from the COMEAP group was published in 1998 and quantified the extent of pre-mature mortality and other health outcomes associated with exposure to air pollutants in Great Britain.[9] The report concluded that the premature deaths associated with PM_{10} levels were 8100, with sulfur dioxide 3500 and ozone 700–12,500 depending on whether or not one assumes a threshold for no adverse effects (in the COMEAP calculations, assuming a threshold of 50ppb led to the lower estimate of mortality). This result was also extremely valuable in informing policy development.

While the UK Air Quality Strategy predecessor documents were in gestation, the EU was similarly formulating a more strategic approach to air quality policy through the so-called air quality Framework Directive (96/62/EC), which was agreed in 1996. This paved the way for 'Daughter Directives' the first of which (99/30/EC) dealt with particles, sulfur dioxide, lead and nitrogen dioxide and was agreed during a UK presidency of the EU and adopted in 1999. Subsequent Directives have dealt with carbon monoxide and benzene ((2000/69/EC), and with ozone (2002/3/EC). The latter Directive on ozone does not contain mandatory limit values like the other Directives, reflecting the transboundary nature of ozone and the fact that a Member State would not necessarily have complete control over the sources of the ozone levels which were measured within its territory. The fourth in the series covers arsenic, nickel, cadmium, mercury and polynuclear aromatic hydrocarbons and like the ozone Directive did not set mandatory limit values but incorporated target values for the pollutants concerned.

In the UK the Air Quality Strategy was revised and updated, and published in 2007. The revision analysed a series of potential measures to improve air quality, and included an extremely comprehensive evidence base covering atmospheric modelling and projections, and a detailed economic analysis of the costs and benefits of the possible measures.[10] The most recent development in Europe has been the agreement of a revised and consolidated Air Quality Directive which brought together for revision the first three 'daughter directives'. The new Directive which was published in May 2008 incorporated some significant developments including the concept of 'exposure reduction' as a new advance in air quality management, foreshadowed in the UK Air Quality Strategy. This is discussed in more detail in the section dealing with future developments.

4 Policy Instruments which Reduce Emissions

Amongst the many policies and regulations on air quality which have been promulgated over the past decades, it is worth distinguishing the more important among them which have actually required the reduction in emissions of air pollutants as opposed to setting standards or broader frameworks.

In the years following the 1956 Clean Air Act, action on sources other than coal has been implemented. In 1966, the Road Vehicle (Construction and Use) Regulations prohibited the emission of 'any smoke, visible vapour, grit, ashes, cinders or oily substance'. Since that time, legislation on motor vehicle emissions has been rather more scientifically based.

Other than the 1956 Clean Air Act, the one series of instruments which has led to the most important improvement in overall public exposure to harmful air pollutants has been the European regulations on motor vehicle emissions. The first EU Directive dated from 1970 (Directive 70/220/EEC) and applied to light duty vehicles. This and subsequent amendments essentially set a cap on emission performance rather than forced large reductions and it was not until the amendment embodied in Directive (88/76/EEC), the so-called 'Luxembourg Agreement', that significant reductions were set in motion. This Directive effectively mandated the use of three-way catalysts (so-called because they reduced levels of all three regulated pollutants, carbon monoxide, hydrocarbons or VOCs, and oxides of nitrogen). A fully functioning catalyst was able to reduce emissions by more than 90% compared to an uncontrolled vehicle. A parallel measure for heavy duty vehicles was agreed in 1988 in Directive 88/77/EC.

A subsequent series of Directives limiting vehicle emissions, covering light and heavy duty vehicles, and extending to particles as well as gaseous emissions, have been agreed since these two Directives and successive standards have become known as the 'Euro X' standards. The 'Luxembourg Agreement' 88/76/EEC is designated Euro 1 and subsequent improvements are now up to Euro 5 and 6, Euro 5 coming into force in September 2009. The parallel series of heavy duty standards are designated with Roman numerals and the most recent level in force is known as Euro IV (Directive 2005/55/EC) and Euro V came into force at the time of writing in October 2008. This set of regulations beginning with the 1987 and 1988 Directives have had the single most important influence on urban air quality in the UK and the rest of Europe in the past two decades, and the most important in the UK since the 1956 Clean Air Act.

Airborne lead concentrations are now extremely low because of the removal of lead from petrol – a prerequisite for the use of catalytic converters. Initially, however, the lead content of petrol was reduced because of concern over its toxic properties. At the end of 1985, the maximum permitted lead content of petrol was more than halved (from 0.4 to 0.15 grammes litre^{-1}) and the airborne lead levels throughout the UK (and EC Europe) more than halved in a matter of weeks. This still probably represents the most abrupt improvement in measured air quality in present times in the UK. Levels reduced to almost zero over a longer period as lead was gradually removed from petrol.

Industrial emissions have also continued to be regulated through a series of significant developments in legislation, drawing on a history of over a hundred years. Recent developments have seen the integration of controls on emissions to air with those to other media in the system of Integrated Pollution Control in the Environmental Protection Act of 1990. This extended and improved the concept of 'Best Practicable Means' which had served industrial pollution control in the UK for over 100 years, and introduced the concept of 'Best Available Techniques'. In the aftermath of EPA 90, the Government created the Environment Agency with a remit of implementing pollution control legislation on emissions to all media. The Agency was created from the former National Rivers Authority and Her Majesty's Inspectorate of Pollution (formerly the Alkali Inspectorate). This system of industrial air pollution control and the parallel system of Local Authority Air Pollution Control have been responsible for the significant reductions in emissions from stationary sources over the past few decades. More recently, the system has developed to incorporate the EC Directive on Integrated Pollution Prevention and Control (96/61/EC). These systems together with the regulation of motor vehicle emissions and reductions in emissions from agricultural emissions will be responsible for the UK meeting its targets under the EC National Emissions Ceilings Directive (2001/81/EC) and the UNECE Gothenburg Protocol.

An evaluation of the air pollution reduction policies was carried out on behalf of Defra in 2004.[11] A thorough analysis was undertaken of the reductions in emissions and concentrations resulting from the policies in the Air Quality Strategy since 1990, chiefly those relating to road transport and the electricity supply industry (ESI), because these are the sectors of the economy where the largest reductions have occurred over that period. The improvements in public health and the economic costs and benefits were also analysed, and care was taken to attribute reductions specifically to air pollution policies, excluding those reductions which would have taken place for other reasons. Space does not permit an exhaustive discussion of the findings of this work, but some conclusions are of interest in the present context. Between the years 1990 and 2001, the study estimated that reductions of sulfur dioxide, nitrogen oxides and PM_{10} from road transport were respectively 96%, 36% and 48% compared with the 'no policies' values, and those from the ESI were 77%, 58% and 78%, respectively. It was further anticipated that by 2010 the corresponding reductions from road transport would be 96%, 69% and 76% and from the ESI 93%, 69% and 93%, respectively. The study concluded that all of the road transport reductions were attributable to air quality policies, but anywhere between $\sim 35\%–100\%$ of the ESI reductions could be the result of air quality policies.

5   Future Developments in Air Quality Management Policies

5.1   *Exposure Reduction*

Recent experience in the UK in formulating measures to attain the objectives in the UK's Air Quality Strategy and European Union limit values, suggested that an air quality management system based *solely* on objectives, limits, or standards expressed as a target concentration to be met in all locations, might be inefficient in terms of improving public health. As emission reduction policies bite and areas in exceedence of a legal limit become smaller, it will generally become necessary to impose more and more costly measures to achieve full compliance. The problem is that the legal framework based on a single mandatory limit value or standard requires precisely this. Cost benefit ratios therefore become increasingly unfavourable as legal compliance is sought by 'chasing hot-spots'. This inefficiency requires a new approach to air quality management, outlined below.

For pollutants with no threshold of effect at the population level such as PM (at the level of current knowledge) it will generally be more beneficial for public health as a whole to reduce concentrations across the whole of an urban area, even where they already meet existing legal limits, than to 'chase hot spots'. The fundamental problem, however, is that legal frameworks based on ambient concentration standards require complete compliance and therefore exert pressures in the wrong direction from what would be required to optimize abatement strategies in terms of improving public health.

The one important benefit of ambient standards or limit values is that compliance ensures a common standard of air quality for all citizens, and this should not be discarded lightly. However, the problems with attaining the limits have been set out above and if limits or standards were to be retained, another criterion would be needed to optimize the system to direct attention to improving public health in some optimal way. Work in the UK has been directed towards the concept of a target for policy framed in terms of reducing 'population exposure' to ambient levels of a pollutant like PM, and expressed as a target to reduce concentrations *averaged over a whole urban area, or region,* by some given percentage over a defined period. This is illustrated in Figure 1[10] showing how the idea is to shift the whole distribution of the exposed

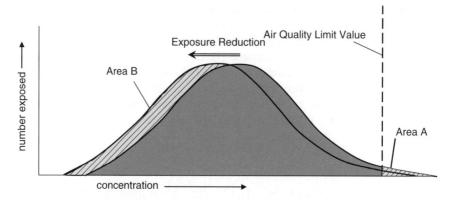

Figure 1   Illustration of the exposure reduction concept.

population to lower concentrations, rather than simply to remove the small area above the standard or limit value in the 'old' system of air quality management.

One could therefore conceive of a system of air quality management which embodies the two measures or targets: an overall limit on ambient levels to ensure some basic level or quality of air which all citizens could experience, embodying the 'environmental justice' concept, and an additional commitment to reduce ambient levels, or exposures, by a given amount even in areas where the ambient limit might already be achieved. To improve the public health gains to be obtained from this new system, it is the latter 'exposure' criterion which should be the driver for policies, not the 'cap' or limit value, or else the inefficient status quo is unchanged. The extent of this 'exposure' reduction could be determined by the balance of costs and benefits. The practicability of such a system would depend crucially on the balance of stringency between the two commitments. To work most efficiently, the limit or standard would ideally be set at a level somewhat less stringent than might otherwise have been the case had the limit been the only driving factor in air quality improvement. This could clearly cause some initial difficulties in adding on an exposure-reduction target to an existing standard or limit value. However, if one is starting from scratch, as the EU was with legislation on $PM_{2.5}$, then the system is potentially workable, and indeed this concept was embodied in the revision of EU legislation which resulted in the Air Quality Directive 2008/50/EC published in May 2008.

In principle, there are issues to resolve around the precise metric for determining the exposure reduction target – should it be in terms simply of measured concentrations or should it be on the basis of some measure of population exposure determined by a combination of measurements and modelling? There are also issues around the spatial area over which the target should apply; is there for example a single target for the whole of a country, or should it be applied in each region? In any new concept such as this one, there is a danger in making the system too complex. For effective policy making, the instruments need to be as simple and straightforward as possible, without losing the essence of the concept. In practice, in the EU Directive, the exposure reduction concept was defined for $PM_{2.5}$ in terms of an Average Exposure Indicator (AEI), defined as the average of annual mean concentrations at urban background locations, reflecting public exposure, over the whole territory of the Member State and averaged over three calendar years. The Directive then required the AEI to be less than $20\,\mu\text{g m}^{-3}$ by 2015 (the 'exposure concentration obligation') and required Member States to take all necessary measures not entailing disproportionate costs to reduce exposure to $PM_{2.5}$ with a view to attaining the 'national exposure reduction target' which was set as a percentage reduction, the size of which is dependent on the initial value of the AEI in a Member State. This is designed so that Member States which start from low values of the AEI do not have to make disproportionately large reductions compared with those starting from much higher values. The concept of the 'cap' referred to above has been incorporated in the Directive as a two-stage limit value for *concentrations* of $PM_{2.5}$ of $25\,\mu\text{g m}^{-3}$ to be met by 2015 (and a non-mandatory target value of the same value to be met by 2010) with an indicative second stage limit value of $20\,\mu\text{g m}^{-3}$ to be achieved by 2020.

Such a system would in principle provide a much more optimal air quality management system than one relying solely on ambient standards. Objections, or counter arguments could be raised that no system does exactly that, and that ambient standards are often also supported by source-related legislation which adopts an approach of continual improvement through a system based on concepts such as 'Best Available Techniques'. While this is true, as long as these legislative systems rely on legally binding ambient standards in hot spot areas, they will still suffer from the inefficiencies described above. Another objection could be that the same efficient result could be achieved by continually tightening the ambient standard over a period of years. This would not necessarily avoid the problem of 'chasing hot spots' and would also impose a considerable bureaucratic and administrative overhead into the system, as well as sacrificing simplicity in implementation. It is also arguable that a system combining ambient standards and an overall emission ceiling for a country or region could fulfil the same purpose. However, this would not have the necessary focus on the improvement of public health. It is possible in principle that the total emission ceilings could be achieved with a disproportionately small improvement in public health, depending on the spatial relationship between the emission reductions and the populations exposed.

It is interesting to note that the exposure reduction approach also embodies a form of environmental justice, although of a different kind from the ambient standards. As long as there are sources of emission in an urban area, then there will always be differences in *exposures* due to dilution and dispersion, even if there is uniformity in compliance with ambient *standards*. If the exposure reduction approach is adopted, and if the reduction amount is required to be the same everywhere, then there will be uniformity in the *improvement* in exposure, in percentage terms, if not in absolute amounts.

The reduction of ambient air levels is becoming increasingly more difficult and costly, yet the evidence from the health effects community is that adverse effects are still present. It is therefore incumbent on air quality managers and policy makers to think creatively and to devise systems to optimize expenditure on air pollution control wherever possible, in order to obtain the most effective protection of public health.

5.2   Links and Co-benefits with Climate Change Policies

The importance of climate change and the need to reduce the emissions of carbon dioxide and other greenhouse gases (GHGs) has increased enormously in the past five years or so. There are very clearly close links with policies to reduce air pollutant emissions since both subjects deal with almost the same set of sources. There will be policies which improve both and will reinforce each other, but there will also be some policies designed to reduce emissions of one set of pollutants but which lead to increases in emissions of the others. The challenge for the policy maker will increasingly be to maximise the benefits from the synergistic policies while minimising the *disbenefits* from those policies which might be in conflict, or which will require the management of 'trade-offs'. A diagrammatic summary of some examples of these policies is given in Figure 2.

Examples of synergistic policies are those which reduce demand for energy or for transportation use. Where these can be implemented through measures such as improved insulation in buildings then their acceptability by the public should not cause problems. Other measures however such as curtailing the public's desire to travel by car need more thought in order to gain wide acceptance. Other so-called 'win-win' policies include the increased use of 'pure' renewables, 'pure' in the sense that they entail zero emission of all pollutants at point of use – both GHGs and air pollutants. Technologies such as wind, solar and tidal power, as well as nuclear, fall into this category. Included in the win-win quadrant of Figure 2 is Carbon Capture and Storage, but the situation is a little more complex than this diagram might imply. Currently, the front runner in CCS technologies in terms of wide availability is post-combustion CCS, but while this technique will probably result in large reductions in sulfur dioxide, it comes with a relatively large fuel penalty, possibly of the order of 30% compared to an uncontrolled (in the carbon dioxide sense) power plant. This in turn could lead to higher emissions of NO_x than an uncontrolled plant, and possibly of ammonia if this gas or amines are used as a solvent to remove the CO_2. Pre-combustion CCS, where the original fuel is gasified to produce a fuel rich in hydrogen with the carbon being removed at the gasification stage, offers more scope for reductions in emissions of NO_x but the reductions may not be very large compared with a non-CCS plant. This technology is probably further away from widespread use than is post-combustion CCS. The furthest away from widespread use is the so-called oxy-fuel technique where the original fuel is gasified and the gas then burned in an atmosphere of almost pure oxygen.

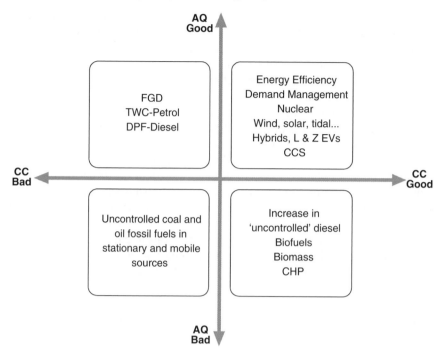

Figure 2   Schematic diagram of the synergies and trade-offs between air quality and climate change policies.

Since the bulk of NO_x emissions in conventional combustion systems arises from atmospheric nitrogen, the NO_x emissions in the oxy-fuel technique, and those of other pollutants, could be very much lower than in other systems.

An important area of trade-off concerns the use of after-treatment technologies to clean up air pollutant emissions from road vehicles and from power stations. The most topical examples concern the fitting of particle filters to light- and heavy-duty vehicles in recent EU legislation. These devices are extremely efficient in removing the emissions of particulate matter from diesel exhaust. This is a major advance in improving air quality as particles are probably the single most important air pollutant affecting human health, but fitting filters comes with a fuel consumption penalty. Estimates of this penalty in percentage terms vary but are in the region of single figures which may seem small, but in the context of challenging carbon reduction targets is significant. In agreeing the Euro 5 and 6 standards for cars, the EU has tacitly accepted the CO_2 penalty in order to achieve the improvements in public health. In some ways however this measure could even be seen as a move towards a win-win position in the sense that if car fleets replace older diesels with filter-equipped diesels then both air quality and climate can benefit. A contribution to the climate benefit also comes from the reduction in particle emissions as black carbon (BC) is a powerful warming agent in the atmosphere with a large radiative forcing. Its effects on global temperatures, however, are modified by its relatively short atmospheric lifetime which is of the order of days. The problem was addressed by Boucher and Reddy[12] who produced an initial analysis using a global climate model, and presented a helpful numerical method for assessing the climate impact of the balance between reduction in BC (beneficial to climate change) and the increase in CO_2.

Another important and live issue concerns the use of biomass (essentially wood) burning as a means of achieving carbon reductions. While the carbon benefits are clear wherever biomass replaces fossil fuel use, the air quality impacts need careful consideration. Biomass burning in large (greater than around 20–50 MW) plant is likely to be efficient and in the larger plant will be subject to tight emission limits so that air quality impacts should be small. The air quality concerns arise from possible widespread use of biomass in small boilers and appliances down to the domestic scale. Inefficient and uncontrolled combustion can lead to significantly higher emissions of particulate matter with implications for public health and for compliance with air quality standards and limits. This is recognised as a problem in many countries in Europe and elsewhere, and a range of measures and approaches have been adopted to address the problem, ranging from public awareness campaigns, financial assistance to home owners to replace their appliances, to tight standards for appliances and the use of the planning process to limit the extent of biomass burning in areas where air quality is a potential problem.

In the long term however, out to 2050 and beyond for example, many countries and organisations are thinking in terms of major reductions in GHG emissions, and the UK recently announced its intention to adopt a target for 2050 of a reduction of 80% in GHGs on a 1990 base, and other countries and organisations are considering similar goals. There will of course be a range of options for achieving these goals, which will have different levels of co-benefits with air quality and public health depending on the strategies adopted. In theory, the 'clean' way of achieving these targets would involve major changes in energy and transport infrastructures to introduce essentially zero emission technologies and techniques – zero emissions of carbon and of air pollutants. The impact on urban air quality of such a hypothetical approach was considered by the present author[13] who showed that levels of particulate matter and other pollutants could be reduced in the long term by around 55% compared with current levels, with the correspondingly significant reductions in adverse effects on human health.

This finding and other studies on the co-benefits of optimised air quality and climate change policies point the way forward for a long term strategy for air quality improvements in developed as well as developing countries. In developed countries like the UK and the rest of the EU, and North America, virtually all the technological solutions developed purely for air quality reasons have now been incorporated into policy instruments. They may not have worked their way into full time use yet – some recently agreed vehicle emission standards have yet to come into force for example. Further advances in technology are now being driven by climate change concerns, and non-technical measures involving behaviour change, consumption patterns, transport choices and planning strategies are also being driven by climate considerations. In terms of strategic policy development therefore, it is becoming clearer that there are significant improvements in air quality, in impacts on public health and on the wider environment, from recognising the importance of climate change and pursuing policies which have the potential to make significant improvements in both problems. This is the challenge for policy makers in addressing the urban environment over the coming decades.

Disclaimer

The views expressed in this paper are those of the author and do not necessarily represent those of the Department of Environment, Food and Rural Affairs.

References

1. *WHO Guidelines for Europe*, 1987.
2. *This Common Inheritance – Britain's Environmental Strategy. Cm 1200*, HMSO, September 1990.
3. *Sustainable Development – The UK Strategy. Cm 2426*, HMSO, January 1994.
4. *A Better Quality of Life – A Strategy for Sustainable Development for the UK. Cm 4345*, HMSO, May 1999.
5. *Improving Air Quality – A discussion paper on Air Quality Standards and Management*, Dept. Environ., March 1994.
6. *Air Quality: Meeting the Challenge – The Government's Strategic Policies for Air Quality Management*, Depart. Environ., Depart. Transport, Welsh Office, DoE(N Ireland), January 1995.
7. *The United Kingdom National Air Quality Strategy, Cm 3587*, The Stationery Office, March 1997.
8. Depart. Health, *Committee on the Medical Effects of Air Pollutants. Non-Biological Particles and Health*, HMSO, London, 1995.
9. Depart. Health, *Committee on the Medical Effects of Air Pollutants. The quantification of the effects of air pollution on health in the United Kingdom*, HMSO, London, 1998.
10. Defra, *The Air Quality Strategy for England, Wales, Scotland and Northern Ireland*, HMSO, London, 2007.
11. AEA Technology, *An Evaluation of the Air Quality Strategy*, January 2005.
12. O. Boucher and M. S. Reddy, *Energy Policy*, 2008, **36**, 193–200.
13. M. L. Williams, *Environ. Sci. Policy*, 2007, **10**, 169–175.

Subject Index